# Cost Estimating Manual for Pipelines and Marine Structures

## Man-Hour Manuals by John S. Page

Estimator's Electrical Man-Hour Manual/
3rd Edition

Estimator's Equipment Installation
Man-Hour Manual/3rd Edition

Estimator's General Construction
Man-Hour Manual/2nd Edition

Estimator's Man-Hour Manual on Heating,
Air Conditioning, Ventilating, and Plumbing/
2nd Edition

Estimator's Piping Man-Hour Manual/5th Edition

### Also by John S. Page

Conceptual Cost Estimating Manual/2nd Edition

---

**John S. Page** has wide experience in cost and labor estimating, having worked for some of the largest construction firms in the world. He has made and assembled numerous types of estimates including lump-sum, hard-priced, and scope, and has conducted many time and method studies in the field and in fabricating shops. Mr. Page has a B.S. in civil engineering from the University of Arkansas and received the Award of Merit from the American Association of Cost Engineers in recognition of outstanding service and cost engineering.

# Cost Estimating Manual for Pipelines and Marine Structures

JOHN S. PAGE

Gulf Professional Publishing

*An Imprint of Elsevier*

This edition reviewed
by the author and
reprinted February 2000.

# Cost Estimating Manual for Pipelines and Marine Structures



Gulf Professional Publishing
An Imprint of Elsevier
Book Division
P.O. Box 2608 ☐ Houston, Texas 77252-2608

**ISBN-13: 978-0-87201-157-1**
**ISBN-10: 0-87201-157-7**

Library of Congress Catalog Card No. 76-40868

Transferred to Digital Printing 2009

## DEDICATION

To all those engaged in
estimating on projects for the
production and transportation
of oil and gas for energy.

# Contents



## Section One—CROSS-COUNTRY PIPELINES

## Section Two—MARSHLAND PIPELINES



## Section Three—ONSHORE AND OFFSHORE SURF-ZONE PIPELINES

## Section Four—OCEAN AND SEA PIPELINES



## Section Five—DREDGING, BLASTING AND JETTING

## Section Six—OFFSHORE STRUCTURES

## Section Seven—OFFSHORE INSTALLATION OF EQUIPMENT AND APPURTENANCES

## Section Eight—MISCELLANEOUS CONSTRUCTION ITEMS



## Section Nine—MISCELLANEOUS ESTIMATING INFORMATION

# Preface

This manual has been compiled to provide time frames, labor crews and equipment spreads to assist the estimator in capsulizing an estimate for the installation of cross-country pipelines, marshland pipelines, nearshore and surf zone pipelines, submerged pipelines, wharfs, jetties, dock facilities, single-point mooring terminals, offshore drilling and production platforms and equipment and appurtenances installed thereon.

The time frames and labor and equipment spreads which appear throughout this manual are the result of many time and method studies conducted under varied conditions and at locations throughout the world; these time frames and labor and equipment spreads reflect a complete, unbiased view of all operations involved.

When one is engaged in compiling an estimate from any information furnished by others, as is the case with this manual, he should view it in an objective light, giving due consideration to the nature of the project at hand and evaluating all items that may affect the productivity of labor and all other elements involved.

## The Human Factor in Estimating

In this high-tech world of sophisticated software packages, including several for labor and cost estimating, you might wonder what a collection of estimating tables offers that a computer program does not. The answer is the *human factor.* In preparing a complete estimate for a heavy industrial project one often confronts 12–18 major accounts, and each account has 5–100 or more sub-accounts, depending on the project and its engineering design. While it would seem that such numerous variables provide the perfect opportunity for computerized algorithmic solution, accurate, cost-effective, realistic estimating is still largely a function of human insight and expertise. Each project has unique aspects that still require the seasoned consideration of an experienced professional, such as general economy, projects supervision, labor relations, job conditions, construction equipment, and weather, to name a few.

Computers are wonderful tools. They can solve problems as no human can, but I do not believe construction estimating is their forté. I have reviewed several construction estimating software packages and have yet to find one that I would completely rely on. Construction estimating is an art, a science, and a craft, and I recommend that it be done by those who understand and appreciate all three of these facets. This manual is intended for those individuals.

*John S. Page*

# Introduction

Any logical system used in the preparation of an estimate for land and sea pipelines and offshore facilities must be based on, and confined to, certain predetermined parameters. Therefore, your particular attention is directed to the following:

No consideration has been given to the dollar value of labor, materials or construction equipment. The dollar cost of labor and material will vary, depending on location and time or schedule of the particular project. Outside or third party rental/purchase of construction equipment will vary, depending on location, availability and the market value at the time the equipment is needed. If the equipment is contractor-owned, its costs will vary depending on its depreciated value, cost of repairs and maintenance, insurance, desired return on investment, etc.

It is of the utmost importance that the correct labor crews and equipment spreads be established and used. No piece of equipment can function properly without its complement of skilled labor to operate and maintain it, and labor cannot produce efficiently without the proper tools and equipment to accomplish the job intended.

Where size or capacity is listed with a piece of equipment in the equipment spreads, it is intended solely as a guide and should not be construed as the only size or capacity that can be used. In actual practice, the project scope and conditions will dictate the size, type and capacity of the equipment to be used. The time frame tables in this manual are based on employing the number and type of units listed in the spread tables.

In keeping with the many time frame tables which appear throughout this manual, the labor crew and equipment spread tables are intended to provide ample labor and equipment for the installation of all items listed. In addition to the General Notes and the notes that appear with the individual tables, consideration should be given to the following when forming and applying these crews and spreads.

1) All labor and equipment spreads can be adjusted upward or downward, depending on project type, size and scope.
2) All equipment should be of the size and type suitable for the work intended.
3) Various crews and equipment spreads, such as derrick barge, tug, crewboat, etc., must be combined to obtain a total working spread. The type and scope of the project will determine the spreads that are to be combined.
4) Offshore construction material and/or cargo barges, in the number and size needed, must be added to the spread at their daily rates for the time they are actually involved at the construction site.

All labor crews listed in this manual are based on open shop operations. If union or closed shop operations are the case, general foremen, various craft foremen, stewards and other craft classifications will have to be given consideration and added to the crews as necessary.

The following items are listed and defined in Section 8, "Miscellaneous Construction Items," and, therefore, no consideration has been given them in the other sections of this book:

1) Mobilization and demobilization of personnel
2) Mobilization and demobilization of equipment
3) Load-out, tie-down and material handling
4) Transportation and freight
5) Camp, family quarters, field office and storage facilities
6) Rest and recreation

Small tools, consumable supplies, fuel, oil and grease have not been considered. These are all items that vary widely with particular projects and should be added to the various spreads as applicable.

Purchase of right-of-way, crop or structure damage or repair or replacement costs, cost of other damaged items, taxes, permits, licenses, fees of all types and duties have not been given consideration in this manual. These can and do vary so greatly that they must be given individual consideration for each location and project and added to the estimate accordingly.

Where work is to be performed in a foreign country, consideration should be given to the forming and use of a mixed crew comprised of at least a portion of craftsmen from that country. Usually this fosters a better relationship with the country, and it is often less expensive than a completely expatriate crew. It should be recognized an *expatriate* craftsman is any craftsman who is not a citizen of the country in which the project is being constructed.

Tables have not been included for overall project management, field management or supervision. The scope of the individual project will govern the need for these personnel, and they must be added to the estimate for a project requiring their services.

## Labor Productivity

Before one thinks in terms of labor dollars for an estimate, many things must be considered, the most important of which are called *productivity efficiency* and *production elements*. Consideration of these two factors is essential if the many labor time frames that follow are to be correctly applied.

After extensive comparison of many projects, we have found that production percentages can be classified into five categories and that production elements can be grouped into six different listings or classifications:

*Production Elements*

1. General Economy
2. Project Supervision
3. Labor Relations
4. Job Conditions
5. Equipment
6. Weather

*Productivity Efficiency Percentages*

| Type | Percentage Range |
|---|---|
| 1. Very Low | 10 Through 40 |
| 2. Low | 41 Through 60 |
| 3. Average | 61 Through 80 |
| 4. Very Good | 81 Through 90 |
| 5. Excellent | 91 Through 100 |

From the above paragraph you may agree that this is true but that the percentage productivity range is too wide to accommodate accuracy. By evaluating each of the six elements and illustrating with an example of each, you can see just how simple it is to arrive at a productivity efficiency percentage.

1. GENERAL ECONOMY

This is simply the state of the nation or the area in which your project is to be developed. Things that should be looked at and evaluated under this category are:

(a) Business Trends and Outlooks
(b) Construction Volume
(c) The Employment Situation

Let us say that after giving due consideration to these items you find them to be very good or excellent. This sounds good but actually it means that your productivity range will be very low. This is due to the fact that with business being excellent the type of supervision and craftsmen that you will have to draw from will be very poor. Because of this it will tend to create bad labor relations between your company and supervision, thus making very unfavorable job conditions. From this it would seem that the general economy of the nation or area sets off a chain reaction to the other five elements. This we have found to be true. On the other hand, let us say that we have evaluated this element and find the general economy to be of a fairly good average. Here we find that the productivity efficiency tends to rise. This is due to the fact that under normal conditions there are enough good supervisors and craftsmen to go around, they are satisfied, thus creating good job conditions.

For our example, to show how a final productivity efficiency percentage can be arrived at, let us say that we are estimating a project in a given area and after careful consideration of this element, we find it to be of a high average. Since it is of a high average, but by no means excellent, we estimate our productivity percentage at seventy-five (75) per cent.

2. PROJECT SUPERVISION

What is the caliber of your supervision? What experience have they had? What can you afford to pay them? What have you to draw from? Things that should be looked at and evaluated under this element are:

(a) Experience
(b) Supply
(c) Pay

Like general economy this too must be carefully analyzed. If business is excellent, the chances are that you will have a poor lot to draw from. If business is normal, you will have a fair chance of obtaining good supervision. The contractor who tries to cut overhead by using cheap supervision usually winds up doing a very poor job. This usually results in a dissatisfied client, a loss of profit and a loss of future work. However, the estimator has no control over this. It must be left to management. All the estimator can do is estimate his projects accordingly.

To follow through with our example, after careful analysis of the three items listed under this element, we find that our supervision will be normal for this type of work and we arrive at an estimated productivity rate of seventy (70) per cent.

3. LABOR CONDITIONS

Have you a good labor relations man in your organization? Are the craftsmen in the area experienced and satisfied? Are there adequate first class craftsmen in the area? Like project supervision things that should be analyzed under this element are:

(a) Experience
(b) Supply
(c) Pay

The area where your project is to be constructed should be checked to see if the proper experienced craftsmen are available locally or will you have to rely on travelers to fill your needs. Can and will your organization pay the prevailing wage rates?

For our example let us say that for a project in a given area we have found our labor relations to be fair but feel that they could be a little better. Since this is the case, we arrive at an efficiency rating of sixty-five (65) per cent for this element.

4. JOB CONDITIONS

What is the scope of work and just what is involved in the project? Is the schedule tight or do you have ample time to complete the work? What is the condition of the site? Is it on land or at sea? If on land is it high and dry or is it low and muddy? If at sea are the waters relatively calm or are they occasioned by storms? What type of operations are involved? What kind of material procurement will you have? There are many items that could be considered here, dependent on the project; however, we feel that the most important of these items are as follows:

(a) Scope of Work
(b) Site Conditions
(c) Material Procurement
(d) Ease of Operations

By careful study and analysis of the plans, specifications and other project information coupled with a site or area visitation you should be able to correctly estimate a productivity efficiency for this item.

For our example, let us say that the project we are estimating allows ample time to complete the project, that the site location is low and muddy, material procurement will be a bit slow and the ease of operation will be normal for the type of work involved. Therefore, after evaluation we estimate a productivity rating of only sixty (60) percent.

5. EQUIPMENT

Do you have ample equipment to do your job? What kind of shape is it in? Will you have good maintenance and repair help? The main items to study under this element are:

(a) Usability
(b) Condition
(c) Maintenance and Repair

This should be the simplest of all elements to analyze. Every estimator should know what type and kind of equipment his company has as well as what kind of mechanical shape it is in.

Let us assume, for our example, that our company equipment is in very good shape, that we have an ample supply to draw from and that we have average mechanics. Since this is the case we estimate a productivity percentage of seventy (70).

6. WEATHER

Check the past weather conditions for the area in which your project is to be located. During the months that you will be constructing what are the weather predictions based on these past reports? The main items to check and analyze here are as follows:

(a) Past Weather Reports
(b) Rain or Snow, Hot or Cold
(c) Storm Frequency

This is one of the worst of all elements to be considered. At best, all you have is a guess. However, by giving due consideration to the items as outlined under this element your guess will at least be based on past occurrences.

For our example, let us assume that the weather is about half good and half bad during the period that our project is to be constructed. We must then assume a productivity range of fifty (50) per cent for this element.

We have now considered and analyzed all six elements and in the examples for each individual element and have arrived at a productivity efficiency percentage. Let us now group these percentages together and arrive at a total percentage:

| Item | Productivity Percentage |
|---|---|
| 1. General Economy | 75 |
| 2. Production Supervision | 70 |
| 3. Labor Relations | 65 |
| 4. Job Conditions | 60 |
| 5. Equipment | 70 |
| 6. Weather | 50 |
| TOTAL | 390 |

Since there are six (6) elements involved we must now divide the total percentage by the number of elements to arrive at an average percentage of productivity:

390 ÷ 6 = 65 percent average productivity efficiency

At this point we must caution the estimator. This example has been included as a guide to show a method that may be used to arrive at a productivity percentage. The preceding elements can and must be considered for each individual project. By so doing, coupled with the proper manpower tables that follow, a good labor value estimate can be properly executed for any place in the world, regardless of its geographical location and whether it be today or twenty years from now.

# Section One

# CROSS-COUNTRY PIPELINES

It is the intent and express purpose of this section to cover as nearly as possible all erection operations, in the form of labor crews and equipment spreads, that may be necessary for the complete installation of several hundred miles of cross-country pipeline.

The following crew and equipment tables do not take into account the fabrication and installation of compressor or pump stations. These are items such as structures, equipment foundations, equipment installations, inner connecting piping, electrical and instrumentation similar in nature to a small process type plant and as such should be separately evaluated.

The manpower listed in the following crew tables is for a single ten (10) hour shift, and the equipment listed in the spread tables is for a single spread. More than one crew and spread may be required, depending on project scope and schedule.

# DETERMINING TOTAL DIRECT DAILY COST OF LABOR AND EQUIPMENT SPREADS

It is assumed that at least one pickup truck in each spread will be equipped with short-wave radio equipment capable of communicating with all required office, storage and individual spread facilities and that each spread will be equipped with walkie-talkie radios for communicating between the various spreads.

Time frames, labor and equipment spreads for river crossings may be found under Section Two, "Marshland Pipelines."

The total direct daily cost of any labor and equipment spread can be determined by the following:

1) Determine the daily rate, including all fringes, of each craft or position and multiply each rate by the number of men required for that position.
2) Determine daily rate of each piece of equipment and multiply each rate by the number of pieces of individual equipment required.
3) Determine the daily cost of fuel, oil and grease required to run the spread and add this cost to the spread.
4) Determine the cost of small tools and consumable supplies required to support the spread for the project, prorate on a daily basis and add this prorated cost to the spread.
5) If quartering and catering are required for personnel, estimate the cost per man day, multiply by the number of men in the crew, and add daily rate for the total crew to the spread.
6) Summarize the total cost of items one (1) through five (5) above to arrive at the total direct cost of the spread per working day.

To determine the daily direct cost of a spread for a non-working or stand-by day simply reduce the working day spread cost by the cost of the oil, fuel and grease and the consumable supplies that normally would be used when working.

# AVERAGE PIPELAY TABLE—UNDERGROUND

**Linear Feet of Pipelay per Ten (10) Hour Day**
**in Trench in Level, Rock-Free Terrain**

| Nominal Pipe Size | PIPE WALL THICKNESS IN INCHES | | | | | |
|---|---|---|---|---|---|---|
| | 0.000 through 0.250 | 0.251 through 0.375 | 0.376 through 0.500 | 0.501 through 0.625 | 0.626 through 0.750 | 0.756 through 1.000 |
| 4 | 10,800 | 10,500 | 10,300 | — | — | — |
| 6 | 9,600 | 9,400 | 9,210 | — | — | — |
| 8 | 9,120 | 8,900 | 8,720 | 8,550 | — | — |
| 10 | 8,900 | 8,760 | 8,580 | 8,410 | — | — |
| 12 | 8,600 | 8,400 | 8,230 | 8,070 | 7,910 | — |
| 14 | 8,400 | 8,230 | 8,160 | 8,000 | 7,840 | — |
| 16 | 8,000 | 7,840 | 7,680 | 7,530 | 7,380 | 7,230 |
| 18 | 7,800 | 7,640 | 7,500 | 7,350 | 7,200 | 7,050 |
| 20 | 7,400 | 7,250 | 7,100 | 6,960 | 6,820 | 6,680 |
| 24 | — | 6,600 | 6,470 | 6,340 | 6,210 | 6,080 |
| 30 | — | 6,000 | 5,880 | 5,760 | 5,640 | 5,530 |
| 36 | — | — | 5,400 | 5,300 | 5,200 | 5,090 |
| 42 | — | — | 4,800 | 4,700 | 4,600 | 4,500 |

The construction labor and equipment spreads that follow in this section should provide for all operations that may be involved in the installation of the above quantities in a ten (10) hour work day in level, rock-free ground.

Productivity will vary where different types of terrain and/or rock are encountered.

See following respective tables and explanations covering productivity factors to be applied for variation in terrain and rock.

See pages xi-xiv for explanation covering productivity factors to be applied for labor.

Above footage is based on installing double random joints of pipe.

# AVERAGE PIPELAY TABLE—ON SUPPORTS

**Linear Feet of Pipelay per Ten (10) Hour Day on Waist-High Supports on Level Ground**

| Nominal Pipe Size | PIPE WALL THICKNESS IN INCHES | | | | | |
|---|---|---|---|---|---|---|
| | 0.000 through 0.250 | 0.251 through 0.375 | 0.376 through 0.500 | 0.501 through 0.625 | 0.626 through 0.750 | 0.751 through 1.000 |
| 4 | 11,320 | 11,100 | 10,880 | — | — | — |
| 6 | 10,080 | 9,880 | 9,680 | — | — | — |
| 8 | 9,560 | 9,370 | 9,190 | 9,010 | — | — |
| 10 | 9,390 | 9,200 | 9,020 | 8,840 | — | — |
| 12 | 8,800 | 8,620 | 8,500 | 8,330 | 8,160 | — |
| 14 | 8,700 | 8,570 | 8,400 | 8,250 | 8,090 | — |
| 16 | 8,040 | 7,880 | 7,730 | 7,580 | 7,430 | 7,280 |
| 18 | 7,800 | 7,650 | 7,500 | 7,350 | 7,200 | 7,060 |
| 20 | 7,550 | 7,400 | 7,250 | 7,100 | 6,960 | 6,820 |
| 24 | — | 6,720 | 6,590 | 6,460 | 6,330 | 6,200 |
| 30 | — | 6,120 | 6,000 | 5,880 | 5,770 | 5,660 |
| 36 | — | — | 5,440 | 5,350 | 5,240 | 5,140 |
| 42 | — | — | 4,840 | 4,740 | 4,650 | 4,560 |

The construction labor and equipment spreads that follow in this section should provide for all operations that may be involved in the installation of the above quantities of pipe in a ten (10) hour work day on waist-high supports on level ground.

Productivity will vary with contour of terrain and accessibility.

See following respective tables and explanations covering productivity factors to be applied for variation in terrain.

See pages xi-xiv for explanation covering productivity factors to be applied for labor.

Above footage is based on installing double random joints of pipe.

# PRODUCTIVITY AND PERCENTAGE EFFICIENCY FACTORS

To correctly apply the average pipelay per day time frames included in the preceding tables consideration must be given to the effects of labor productivity and terrain and rock factors.

Since the effect of these items has a decisive bearing on the actual linear feet of pipelay one might expect to complete in a day, we have included the following percentage efficiency tables covering these items and an example of their application.

Productivity factors to be applied to linear feet of pipelay per day for labor are as follows:

LABOR PRODUCTIVITY FACTORS

| Type of Productivity | Percentage Range |
|---|---|
| Excellent | 0.901 through 1.000 |
| Very good | 0.801 through 0.900 |
| Average | 0.601 through 0.800 |
| Low | 0.401 through 0.600 |
| Very low | 0.001 through 0.400 |

The following production elements should be given consideration for the application of a labor productivity percentage.

1. General Economy
2. Project Supervision
3. Labor Conditions
4. Job Conditions
5. Equipment
6. Weather

For a full description of above elements and an example of obtaining labor productivity percentage see pages xi-xiv.

Productivity factors for terrain to be applied to linear feet of pipelay per day are as follows:

TERRAIN PRODUCTIVITY FACTORS

| Description of Terrain | Productivity Factor |
|---|---|
| Level | 1.000 |
| Slightly Rolling | 0.950 |
| Rolling | 0.688 |
| Rough | 0.500 |
| Very Rough | 0.300 |

*Level:* Smooth and level, no hills.
*Slightly Rolling:* Level with minor contour changes and small hills.
*Rolling:* Constant contour changes and small hills.
*Rough:* Partially level with minor contour changes, slues and valleys.
*Very Rough:* Constant contour changes, major hills, slues and valleys.

Productivity factors to be applied for rock to linear feet of pipelay per day are as follows:

ROCK PRODUCTIVITY FACTORS

| Percent of Rock | Productivity Factor |
|---|---|
| 0 | 1.000 |
| 0 to 25 | 0.840 |
| 25 to 40 | 0.720 |
| 40 to 55 | 0.627 |
| 55 to 70 | 0.547 |
| 70 to 85 | 0.453 |
| 85 to 100 | 0.373 |

*Percent*

0: No rock—Few small pebbles.
0 to 25: Minor small rocks.
25 to 40: Major rocks and some small boulders.
40 to 55: Major rocks and some small and large boulders.
55 to 70: Minor small and large boulders with frequent occurrence.
70 to 85: Large boulders with frequent occurrence.
85 to 100: Large boulders with frequent to almost continuous occurrence.

# EXAMPLE—APPLICATION OF FACTORS

With such a wide productivity range between various labor, terrain and rock factors, one might think that it is rather difficult to determine just how many feet of pipe one should expect to complete in a day. We know of no fool-proof method for determining this. However, we offer the following example of a method that may be used in the application of the previous listed labor, terrain and rock factor tables. We feel that in so correctly considering and applying these factors one will be very competitive and obtain his share of work at a profit.

For our example we assume that we have four hundred (400) miles of 30-inch by 3/8-inch wall pipeline to install under the following conditions and that we will be working a ten (10) hour day.

From the preceding pipelay table we determine that six thousand (6000) linear feet of 30-inch by 3/8-inch wall pipe can be laid in a trench in level, rock-free ground during a ten (10) hour work period.

From the labor analyzation table (see pages xi-xiv), it has been determined that our labor productivity is equal to sixty-five (65) percent.

Let us assume that the entire pipeline is not to be laid in total level rock-free ground and that the following terrain and rock factors must be applied.

EXAMPLE TERRAIN FACTORS

| Description | Terrain Factor | Miles of Pipe | Use Factor |
|---|---|---|---|
| Level | 1.000 | 100 | 100.0 |
| Slightly Rolling | 0.950 | 105 | 99.8 |
| Rolling | 0.688 | 115 | 79.1 |
| Rough | 0.500 | 50 | 25.0 |
| Very Rough | 0.300 | 30 | 9.0 |
| TOTAL | — | 400 | 312.9 |

$$\text{Terrain Factor} = \frac{312.9}{400} = 0.782$$

# EXAMPLE—APPLICATION OF FACTORS (continued)

EXAMPLE ROCK FACTORS

| Percent of Rock | Rock Factor | Miles of Pipe | Use Factor |
|---|---|---|---|
| 0 | 1.000 | 90 | 90.0 |
| 0 to 25 | 0.840 | 190 | 159.6 |
| 25 to 40 | 0.720 | 15 | 10.8 |
| 40 to 55 | 0.627 | 55 | 34.5 |
| 55 to 70 | 0.547 | 50 | 27.4 |
| TOTAL | — | 400 | 322.3 |

$$\text{Rock Factor} = \frac{322.3}{400} = 0.806$$

Based on the preceding assumptions, the average rate of pipelay per ten (10) hour day is equal to:

$$\text{Linear feet per day} \times \text{labor factor} \times \text{terrain factor} \times \text{rock factor}$$

or

$$6000 \text{ L.F.} \times 0.65 \times 0.782 \times 0.806 = 2458 \text{ L.F. per day}$$

Piping on supports can be estimated in the same manner using factors which may apply.

The direct installation cost, in dollar value, per linear foot of pipe, can be obtained by the following formula:

$$\text{Direct cost per foot} = \frac{\text{Cost per day of spreads}}{\text{Linear feet per day of lay}}$$

# CLEARING AND GRADING RIGHT-OF-WAY

## General Notes

In order to arrive at a total time frame for the following labor crew and equipment spread the following items, as they may apply to a given project, should be given consideration:

1) Removal of trees, brush and stumps.
2) Grubbing and removal of stumps that are in the way of the ditch.
3) Disposal of all debris, including method of disposal and length of haul.
4) Clearing area spoil a sufficient distance from the ditch line so that the spoil-bank from the ditching operations will not fall in any foreign material that might become mixed with the excavated spoil.
5) Cutting of merchantable timber into standard lengths and stacked along the right-of-way for disposition by others if specifically required by the right-of-way agreement.
6) Providing temporary walks, passageways, fences or other structures so as not to interfere with traffic.
7) Providing sufficient and proper lighting where required.
8) Providing guards where required.
9) Preserving all trees, shrubs, hedges and lawns where required.
10) Grading irregularities where required.
11) Preserving topsoil for replacement, through all cultivated or improved fields and pastures, to its orginal position.
12) Proper grading of the terrain so as to allow passage of loaded trucks and equipment hauling materials and so ditching operations can be properly performed.
13) Protecting and preserving existing drainage facilities.
14) Protecting any existing structures or pipelines.
15) Protecting any telephone or utility lines and keeping them in service.
16) Cutting through fences and hedges where required and replacing these when necessary.
17) Installing gates and fencing where required.

# CLEAR AND GRADE RIGHT-OF-WAY

**Labor Crew**

| Personnel Description | NUMBER OF MEN FOR | | | | | | | | | | | |
|---|---|---|---|---|---|---|---|---|---|---|---|---|
| | 50 Linear Feet Width | | | | 80 Linear Feet Width | | | | 100 Linear Feet Width | | | |
| | L | M | MH | H | L | M | MH | H | L | M | MH | H |
| Foreman .... | 1 | 1 | 2 | 2 | 1 | 2 | 3 | 3 | 2 | 3 | 4 | 4 |
| Operator .... | 1 | 1 | 2 | 2 | 1 | 2 | 3 | 3 | 2 | 4 | 5 | 5 |
| Mechanic ... | 1 | 1 | 1 | 1 | 1 | 1 | 1 | 1 | 1 | 1 | 1 | 1 |
| Swamper .... | 1 | 1 | 2 | 2 | 1 | 2 | 3 | 3 | 2 | 4 | 5 | 5 |
| Truck Driver | 2 | 2 | 2 | 2 | 2 | 2 | 2 | 2 | 2 | 2 | 3 | 3 |
| Laborer ..... | 10 | 15 | 20 | 30 | 15 | 25 | 30 | 40 | 20 | 35 | 40 | 50 |
| Total Crew .. | 16 | 21 | 29 | 39 | 21 | 34 | 42 | 52 | 29 | 49 | 58 | 68 |

Above total crew should be ample for clearing and grading one (1) mile of right-of-way per ten (10) hour day for the width and conditions outlined.

Crew spread includes cutting, stacking or piling, loading and hauling a round-trip distance of two (2) miles.

If burning is necessary or permitted, substitute fire tenders for dump truck drivers. See Clearing and Grading equipment spread for number of dump trucks.

*Code Description:*

L = Light—light brush and grass, no trees.
M = Medium—considerable brush of larger size.
MH = Medium Heavy—large brush and small trees.
H = Heavy—much small brush, many small trees and occasional large trees.

# CLEARING AND GRADING RIGHT-OF-WAY

## Equipment Spread

| Equipment Description | NUMBER OF UNITS FOR | | | | | | | | | | | |
|---|---|---|---|---|---|---|---|---|---|---|---|---|
| | 50 Linear Feet Width | | | | 80 Linear Feet Width | | | | 100 Linear Feet Width | | | |
| | L | M | MH | H | L | M | MH | H | L | M | MH | H |
| D8 Tractor W/Dozer ... | 1 | 1 | 1 | 1 | 1 | 1 | 2 | 2 | 1 | 2 | 3 | 3 |
| D7 Tractor W/Dozer ... | 0 | 0 | 1 | 1 | 1 | 1 | 1 | 1 | 1 | 1 | 2 | 2 |
| Truck—2½ Ton Dump .. | 2 | 2 | 2 | 2 | 2 | 2 | 2 | 2 | 2 | 2 | 3 | 3 |
| Truck— Pick-up .... | 1 | 1 | 2 | 2 | 1 | 2 | 3 | 3 | 2 | 3 | 4 | 4 |
| Ripper or Brushrake . | 1 | 1 | 1 | 1 | 1 | 1 | 1 | 1 | 1 | 2 | 2 | 2 |

Above equipment spread should be ample for clearing and grading one (1) mile of right-of-way per ten (10) hour day for the width and conditions as are outlined.

Haul trucks are based on round trip haul of two (2) miles. If brush and trees are to be burned on site, omit above dump trucks.

Small tools such as saws, axes, etc., must be added as required for the individual job.

*Code Description:*

L = Light—light brush and grass, no trees.
M = Medium—considerable brush of larger size.
MH = Medium Heavy—large brush and small trees.
H = Heavy—much small brush, many small trees and occasional large trees.

# PIPELINE LAY-OUT

## Labor Crew

| Personnel Classification | Number of Men |
|---|---|
| Engineer-Surveyor | 2 |
| Rodman | 2 |
| Chainman | 2 |
| Laborer (Stake Driver) | 2 |
| Truck Driver | 1 |
| Total Crew | 9 |

## Equipment and Tool Spread

| Equipment and Tool Description | Units Each |
|---|---|
| Transit | 2 |
| Flagpole | 2 |
| Tape Chains | 2 |
| Ten Pound Sledge Hammer | 2 |
| Hand Brush Hog | 2 |
| Axe | 2 |
| Pick-up Truck | 1 |

Above labor crew, equipment and tools should be ample for surveying and staking out one (1) mile of pipe centerline per ten (10) hour day.

Trench bottom grades can be shot with transit; however, it may be desirable to accomplish this with the use of a level. If level is to be used, add this to above equipment spread.

# UNLOADING, HANDLING, HAULING AND STRINGING PIPE

## General Notes

In estimating the time requirements for the following labor crew and equipment spread, consideration should be given to the following items should they be related to the particular project at hand:

1) Obtain hauling permits as may be required by federal, state and local laws.
2) Offload pipe, valves, fittings, pipe materials, casings, etc., at point of delivery if required and haul to storage facilities.
3) Secure storage sites.
4) Unload and store on racks if required.
5) Check material for damage when received.
6) Load, haul and string pipe and materials along right-of-way.
7) When handling pre-coated pipe, protect coating by use of padded chains or slings.

# UNLOADING, HANDLING, HAULING AND STRINGING PIPE

### Labor Crew

| Personnel Classification | MANPOWER FOR PIPE SIZES | | |
|---|---|---|---|
| | 4″ to 14″ | 16″ to 24″ | 30″ to 42″ |
| Foreman | 2 | 2 | 2 |
| Operator | 3 | 3 | 3 |
| Oiler | 3 | 3 | 3 |
| Truck Driver | 4 | 5 | 6 |
| Hooker | 2 | 2 | 2 |
| Laborer | 8 | 8 | 10 |
| Total Crew | 22 | 23 | 26 |

### Equipment Spread

| Equipment Description | EQUIPMENT EACH FOR PIPE SIZES | | |
|---|---|---|---|
| | 4″ to 14″ | 16″ to 24″ | 30″ to 42″ |
| Crane | 1 | 1 | 1 |
| D7 Toe Tractor | 0 | 1 | 1 |
| D6 Toe Tractor | 1 | 0 | 0 |
| D7 Side Boom | 0 | 1 | 1 |
| D6 Side Boom | 1 | 0 | 0 |
| Truck W/Pipe Trailer | 4 | 5 | 6 |
| Truck—Pick-up | 2 | 2 | 2 |

Above labor crew and equipment spread is ample for handling, hauling, unloading and stringing pipe along a right-of-way that is accessible to vehicles involved.

Length of haul and size and quantity of pipe will determine time for which above labor crew and equipment spread will be required.

Allow enough time for appropriate crew and equipment to maintain a safe lead over the pipe installation crew.

If blasting is required, do not string pipe until blasting is completed.

Above trucks should be equipped with shortwave radios.

# DITCHING AND TRENCHING

## General Notes

In determining a time frame for application of the following labor crew and equipment spread the following should be given consideration should they apply to the particular project:

1) Ditching or trenching for buried pipelines should be in accordance with the following table of minimum width and coverage for all soil formations.

| Nominal Pipe Size Inches | Minimum Width Inches | Normal Minimum Coverage Inches |
|---|---|---|
| 4 | 22 | 30 |
| 6 | 26 | 30 |
| 8 | 26 | 30 |
| 10 | 26 | 30 |
| 12 | 30 | 30 |
| 14 | 32 | 30 |
| 16 | 36 | 30 |
| 18 | 38 | 30 |
| 20 | 40 | 30 |
| 24 | 44 | 30 |
| 30 | 50 | 30 |
| 36 | 52 | 36 |
| 42 | 58 | 36 |

2) In rock cut ditches at least six (6) inches wider.
3) If dirt-filled benches are used, ditch should be excavated deeper to obtain proper coverage.
4) Trench should be excavated to greater depth when required for proper installation of the pipe where the topography of the country warrants same.
5) Repair any damage to and maintain existing natural or other drainage facilities.
6) Do not open ditch too far in advance of pipelay crew.
7) Obtain permits for blasting.
8) When blasting, use extreme caution and protection.
9) Clean up of blasted rock to prevent damage to coated pipe.

# DITCHING AND TRENCHING

## Labor Crew

| Personnel Description | Number of Men Required |
|---|---|
| Foreman | 2 |
| Operator | 15 |
| Oiler | 13 |
| Swamper | 13 |
| Grader | 13 |
| Operator—Jackhammer | 2 |
| Operator—Drills | 4 |
| Powder Man | 4 |
| Truck Driver | 6 |
| Laborer | 10 |
| Total Crew | 82 |

Manpower will vary depending on length, width and type of excavation and type of terrain and soil conditions.

The above labor crew is based on conditions outlined in the example of application factors in the front of this section and should be maximum required to open ditch far enough in advance of the pipelaying crews so as not to delay the pipelay operations.

If demolition of rocks and boulders is not necessary, or if no blasting is required, delete the craftsmen for these operations from the above crew.

# DITCHING AND TRENCHING

## Equipment Spread

| Equipment Description | Number of Units Required |
|---|---|
| Trenching Machine | 2 |
| Dragline ½ Cy. to 1 Cy. | 2 |
| Backhoe ⅝ Cy. to 1 Cy. | 5 |
| Tractor W/Dozer D6 or D7 | 1 |
| Tractor W/Tow D6 or D7 | 1 |
| Side Boom D6 or D7 | 3 |
| Air Compressor 365 to 500 CFM. | 2 |
| Jackhammer—Light to Heavy | 2 |
| Drills—Wagon or Train 4″ to 5½″ | 4 |
| Water Pumps 2″ to 6″ | 2 |
| Trucks—Dump 5 to 12 Cy. | 6 |
| Trucks—Pick-up ½ to ¾ Ton | 6 |

Size of above equipment may vary depending on length, width and type of excavation and type of terrain and soil conditions.

The above equipment spread is based on conditions outlined in the example of application factors in the front of this section and should be maximum required to open ditch far enough in advance of the pipelaying crews so as not to delay the pipelay operations.

If demolition of rocks and boulders is not necessary, or if no blasting is required, delete the equipment for these operations from the above spread.

# BENDING OPERATIONS

## General Notes

In estimating the time requirements for bending operations, consideration should be given to the following items:

1) Bending machine should be of proper size and design.
2) All bends should be cold bends without wrinkles.
3) Curvature of bend should be distributed uniformly throughout as great a length of the pipe as possible.
4) Bends should be used for difference in elevation of the surface of the ground as well as for alignment.
5) Bends should not exceed one and one-half degrees (1½°) per diameter length of pipe.
6) Bend should be made no nearer than six (6) feet from the end of the joint of pipe. When pipe is double-jointed before bending, bends should not be closer to the weld than one (1) pipe diameter. If, however, it should be necessary to bend pipe through a circumferential weld, the weld should be x-rayed after bending.

# BENDING OPERATIONS

## Labor Crew

| Personnel Classification | Number of Men Each |
|---|---|
| Foreman | 1 |
| Engineer | 1 |
| Engineer Helper | 1 |
| Operator—Tractor | 2 |
| Operator—Bending Machine | 1 |
| Operator—Bending Mandrel | 1 |
| Oiler | 2 |
| Pipefitter | 2 |
| Laborer | 4 |
| X-Ray Technician | 1 |
| Total Crew | 16 |

## Equipment Spread

| Equipment Description | Units Each |
|---|---|
| Pipe Bending Machine | 1 |
| Pipe Bending Mandrel | 1 |
| Dent Remover | 1 |
| Air Compressor 365 C.F.M. | 1 |
| Tractor W/Side Boom D6 or D7 | 2 |
| Transit | 1 |
| Truck—Pick-up | 1 |
| X-Ray Equipment | 1 |

Above labor crew and equipment spread is ample to carry out bending operations of all sizes.

More than one spread may be desired or above spread may be decreased depending on number and size of bends and their locations.

Size of pipe will dictate size of bending equipment to be used.

# ALIGNING AND WELDING OPERATIONS

## General Notes

In keeping with the average pipelay per day table, consideration should be given to the following items for aligning and welding operations:

1) The labor crew and equipment spread tables that follow are based on welds being made by the manual shielded metal-arc method.
2) Welding machines should be of the size and type suitable for the work.
3) Welding machines are to be operated within the amperage and voltage ranges recommended for each size and type of electrode.
4) Remove all foreign matter from the beveled ends before welding. Clean with hand or power tools.
5) Pipe ends must have proper bevels for the work intended.
6) Abutting pipe ends should be aligned so as to minimize the offset between pipe surface. Hammering of pipe to obtain proper line-up should be held to a minimum.
7) Internal line-up clamps should be used where possible. Use external line-up clamps when it is impractical or impossible to use an internal line-up clamp.
8) If roll welding is used, add skids or structural framework and required roller dollies to the equipment spread.
9) Use minimum of root bead, filler bead and finish bead.
10) Brush and clean the finished weld.
11) Only qualified welders should be used.
12) Where required apply heat treatment—preheating and postheating.
13) Apply radiographic inspection of welds where required.

# ALIGNING AND WELDING OPERATIONS

**Labor Crew**

| Personnel Description | MEN EACH FOR PIPE SIZES | | |
|---|---|---|---|
| | 4″ to 14″ | 16″ to 24″ | 30″ to 42″ |
| Foreman | 1 | 2 | 2 |
| Welder—Stringer | 2 | 3 | 4 |
| Welder—Line | 12 | 15 | 19 |
| Welder—Helper | 15 | 18 | 23 |
| Clamp Man | 2 | 2 | 2 |
| Stabber | 1 | 1 | 1 |
| Spacer | 2 | 2 | 2 |
| Buffer | 2 | 2 | 2 |
| Operator | 3 | 5 | 6 |
| Oiler | 3 | 3 | 3 |
| Swamper | 1 | 2 | 2 |
| Truck Driver | 4 | 5 | 6 |
| Laborer | 6 | 8 | 10 |
| X-Ray Technician | 1 | 1 | 1 |
| Total Crew | 55 | 69 | 83 |

Above labor crew is the maximum that should be required for aligning and welding of pipe as outlined in the average pipelay table and the example application of factors in the front of this section. See pages 4 through 9.

Above manpower is based on aligning and welding several hundred miles of pipeline. Above crew should be adjusted downward for smaller projects in length if the schedule allows ample time to install same.

# ALIGNING AND WELDING OPERATIONS

**Equipment Spread**

| Equipment Description | NUMBER OF UNITS FOR PIPE SIZES | | |
|---|---|---|---|
| | 4″ to 14″ | 16″ to 24″ | 30″ to 42″ |
| 300 Amp Welding Machine | 12 | 20 | 25 |
| Beveling and Cutting Rig | 1 | 1 | 1 |
| D7 Side Boom | 2 | 2 | 2 |
| D6 Tow Tractor | 1 | 3 | 4 |
| Buffing Outfit | 2 | 2 | 2 |
| Line-Up Clamps | 2 | 2 | 2 |
| Truck—2 Ton Winch | 2 | 2 | 3 |
| Truck—5 Ton | 1 | 1 | 1 |
| Truck—Pick-up | 1 | 2 | 2 |
| Carryall or Bus | 1 | 2 | 2 |
| X-Ray Equipment Set-Up | 1 | 1 | 1 |

Above equipment spread is the maximum that should be required for aligning and welding of pipe as outlined in the average pipelay table and the example application of factors, in the front of this section. See pages 4 through 9.

Above equipment spread is based on aligning and welding several hundred miles of pipeline. Above spread should be adjusted downward for smaller projects if the schedule allows ample time to install same.

# CLEANING, PRIMING, COATING AND WRAPPING OPERATIONS

## General Notes

Consideration should be given to the following items when applying the time requirements for the labor crews and equipment spreads on pages 25 through 28:

1) Clean and remove all dirt, mill scale, rust, welding scale and all other foreign material.
2) Some or all of the following steps or applications may be required in applying the coating and wrapping of the pipe or joints and therefore should be given consideration:
    - i) Priming
    - ii) Enamel Finish
    - iii) Fiberglass Wrap
    - iv) Enamel Coat No. 2
    - v) Asbestos Felt Wrap
    - vi) Fiberglass Wrap No. 2
    - vii) Enamel Coat No. 3
    - viii) Kraft Wrapping
    - ix) Unbonded Wrap
    - x) Whitewash
    - xi) Wood Lagging
3) Handle coated pipe with equipment designed to prevent damage to the coating at all times.
4) Where pipe is to be laid in trench excavated in rock, pad trench to prevent damage to pipe coating.
5) Check all coatings for holidays with a high voltage electric holiday detector.

# CLEANING, PRIMING, COATING AND WRAPPING OPERATIONS

**Labor Crew**

| Personnel Description | MEN EACH FOR PIPE SIZES | | |
|---|---|---|---|
| | 4″ to 14″ | 16″ to 24″ | 30″ to 42″ |
| Foreman | 1 | 1 | 1 |
| Operator—Equipment | 3 | 4 | 4 |
| Operator—Cleaning Machine | 1 | 1 | 1 |
| Operator—Coating Machine | 1 | 1 | 1 |
| Operator—Dope Pot | 1 | 1 | 1 |
| Pot Fireman | 1 | 1 | 1 |
| Oiler | 3 | 4 | 4 |
| Paper Lathers | 2 | 2 | 2 |
| Swamper | 3 | 4 | 4 |
| Truck Driver | 1 | 2 | 2 |
| Laborer | 10 | 15 | 20 |
| Total Crew | 27 | 36 | 41 |

Above crew is ample for cleaning, priming, coating and wrapping several hundred miles of pipeline of the sizes listed based on the linear feet per day as outlined in the average pipelay table and further explained in the example application of factors in the front of this section. See pages 4 through 9.

Part of the above manpower can be adjusted downward for smaller projects, assuming the schedule allows ample time to install same.

If pipe is to be received pre-coated and wrapped, see table on page 27 for field cleaning, priming, coating and wrapping of joints only.

# CLEANING, PRIMING, COATING AND WRAPPING OPERATIONS

### Equipment Spread

| Equipment Description | NUMBER OF UNITS FOR PIPE SIZES | | |
|---|---|---|---|
| | 4″ to 14″ | 16″ to 24″ | 30″ to 42″ |
| Cleaning Machine | 1 | 1 | 1 |
| Coating and Wrapping Machine | 1 | 1 | 1 |
| D7 Side Boom | 2 | 3 | 3 |
| D6 Tow Tractor | 1 | 1 | 1 |
| Holiday Detector | 1 | 1 | 1 |
| 6 KW Generator | 2 | 2 | 2 |
| Truck—3 Ton | 1 | 1 | 1 |
| Truck—2 Ton Float | 0 | 1 | 1 |
| Truck—Pick-up | 1 | 1 | 1 |

Above equipment spread is the maximum that should be required for cleaning, priming, coating and wrapping several hundred miles of pipeline of the sizes listed based on the linear feet per day as outlined in the average pipelay table and further explained in the example application of factors, in the front of this section. See pages 4 through 9.

Part of the above equipment can be adjusted downward for smaller projects in length assuming the schedule allows ample time to install same.

Sizing of various pieces of equipment listed above is to be determined by the size of pipe that is to be installed.

If piping is to be received pre-coated and wrapped, see table on page 27 for equipment required to field clean, prime, coat and wrap joints only.

# TOTAL FIELD ONSHORE CLEANING, PRIMING, COATING AND WRAPPING OPERATIONS OF JOINTS ONLY

**Labor Crew**

| Personnel Description | Number of Men Required |
|---|---|
| Foreman | 1 |
| Operator | 1 |
| Oiler | 1 |
| Buffer | 2 |
| Dye, Enamel and Wrappers | 4 |
| Pot Fireman | 1 |
| Laborer | 2 |
| Truck Driver | 1 |
| Total Crew | 13 |

**Equipment Spread**

| Equipment Description | Number of Units Required |
|---|---|
| D7 Side Boom | 1 |
| 3 Bbl. Tar Kettle | 1 |
| Holiday Detector | 1 |
| 6 KW Generator | 2 |
| Buffer | 3 |
| Pick-up Truck | 1 |

The above labor crew and equipment spread is capable of cleaning, priming, coating and wrapping, of pipe joints only, for the quantities of pipelay as are listed in the average per day pipelay table on pages 4 and 5.

If total pipeline is to be field coated and wrapped, see tables on pages 25 and 26.

# SANDBLAST AND PAINT PIPE

## Labor Crew

| Personnel Description | Number of Men Required |
|---|---|
| Foreman | 1 |
| Blasters and Painters | 3 |
| Pot Man or Helper | 4 |
| Operator | 1 |
| Truck Driver | 1 |
| Total Crew | 10 |

## Equipment Spread

| Equipment Description | Number of Units Required |
|---|---|
| Sand Pot and Blast Equipment | 1 |
| Paint Spray Equipment | 1 |
| Air Compressor—600 CFM | 1 |
| Truck—Pick-up | 1 |

The above labor crew and equipment spread is ample for cleaning, by blasting, and painting of pipe installed on supports or racks and should be capable of blasting twelve hundred (1200) square feet and painting one thousand (1000) square feet of pipeline per ten (10) hour day.

# LOWERING PIPE IN TRENCH OPERATION

## General Notes

Consideration should be given to the following operations involved in lowering pipe into a trench:

1) Remove rocks or other objects that could damage pipe from trench.
2) Allow slack in line to properly fit ditch.
3) Use only canvas slings to lower pipe into trench.
4) Use holiday detector behind the last lowering in belt at all times.
5) Where it is necessary to pull or drag sections of pipe, protect coating.
6) Keep open ends of pipe closed at all times with "night-caps."
7) Should dents, buckles or defects be found in pipe replace same.

# LOWERING PIPE IN TRENCH OPERATION

**Labor Crew**

| Personnel Description | MEN EACH FOR PIPE SIZES | | |
|---|---|---|---|
| | 4″ to 14″ | 16″ to 24″ | 30″ to 42″ |
| Foreman | 1 | 2 | 2 |
| Operator | 3 | 4 | 5 |
| Oiler | 1 | 1 | 1 |
| Swamper | 3 | 4 | 5 |
| Welder | 1 | 2 | 2 |
| Welder Helper | 1 | 2 | 2 |
| Truck Driver | 1 | 2 | 2 |
| Laborers | 8 | 10 | 12 |
| Total Crew | 19 | 27 | 31 |

Above labor crew is ample for lowering in several hundred miles of pipeline, into pre-excavated trench, of the sizes listed above based on the linear feet per day as outlined in the average pipelay table, and further explained in the example application of factors in the front of this section. See pages 4 through 9.

Part of the above manpower can be adjusted downward for smaller projects, assuming the schedule allows ample time to install same.

# LOWERING PIPE IN TRENCH OPERATION

**Equipment Spread**

| Equipment Description | NUMBER OF UNITS FOR PIPE SIZES | | |
|---|---|---|---|
| | 4″ to 14″ | 16″ to 24″ | 30″ to 42″ |
| Side Boom D7 or D8 | 2 | 3 | 4 |
| Dragline—½ to ¾ Cy. | 1 | 1 | 1 |
| Welding Machine | 1 | 2 | 2 |
| Water Pump | 1 | 1 | 1 |
| Truck—2 Ton | 1 | 2 | 2 |
| Truck—Pick-up | 1 | 2 | 2 |

Above equipment spread is maximum that should be required for lowering in several hundred miles of pipeline, into pre-excavated trench, of the sizes listed above based on the linear feet per day as outlined in the average pipelay table and further explained in the example application of factors in the front of this section. See pages 4 through 9.

Part of the above equipment can be adjusted downward for smaller projects, assuming the schedule allows ample time to install same.

# VALVE INSTALLATION

## Labor Crew

| Personnel Description | Number of Men Required |
|---|---|
| Foreman | 1 |
| Pipefitter—Welder | 6 |
| Pipefitter—Helper | 6 |
| Operator | 1 |
| Oiler | 1 |
| Laborer | 4 |
| Truck Driver | 1 |
| Total Crew | 20 |

## Equipment Spread

| Equipment Description | Number of Units Required |
|---|---|
| Light Crane | 1 |
| Welding Machine 300 Amp. | 1 |
| Generator | 1 |
| Flatbed Truck—2 Ton | 1 |
| Pick-up Truck | 1 |

The above labor crew and equipment spread should be ample for the installation of valves in pipelines.

Number of valves to be installed, valve size, type and location will govern the actual size of crew and equipment spread required.

See page 33 for average hourly time requirements for valve installation.

# VALVE INSTALLATION

**Average Hourly Time Requirements**

| Nominal Pipe Size | FLANGED AND BOLTED VALVES | | | | |
|---|---|---|---|---|---|
| | SERVICE PRESSURE RATING | | | | |
| | 150 Lb. | 300-400 Lb. | 600-900 Lb. | 1500 Lb. | 2500 Lb. |
| 8 | 17. | 19. | 25. | 35. | 40. |
| 10 | 19. | 24. | 34. | 44. | 50. |
| 12 | 27. | 29. | 42. | 55. | 63. |
| 14 | 31. | 35. | 49. | 63. | 73. |
| 16 | 36. | 40. | 56. | 72. | 85. |
| 18 | 41. | 46. | 64. | 85. | 98. |
| 20 | 48. | 54. | 73. | 108. | 110. |
| 22 | 53. | 60. | 83. | 111. | 114. |
| 24 | 58. | 65. | 92. | 121. | 124. |
| 26 | 65. | 71. | 100. | 123. | — |
| 30 | 79. | 86. | 119. | 144. | — |
| 36 | 100. | 108. | 148. | 177. | — |
| 42 | 121. | 130. | 176. | 211. | — |
| | WELD END VALVES | | | | |
| 8 | 8. | 9. | 10. | 11. | 13. |
| 10 | 10. | 10. | 13. | 15. | 17. |
| 12 | 13. | 13. | 17. | 19. | 23. |
| 14 | 15. | 16. | 21. | 24. | 31. |
| 16 | 19. | 20. | 26. | 30. | 38. |
| 18 | 24. | 25. | 33. | 38. | 49. |
| 20 | 27. | 28. | 40. | 46. | 57. |
| 22 | 31. | 32. | 47. | 53. | 65. |
| 24 | 35. | 37. | 54. | 61. | 72. |
| 26 | 40. | 42. | 69. | 75. | 93. |
| 30 | 49. | 51. | 82. | 89. | 109. |
| 36 | 69. | 72. | 113. | 125. | 141. |
| 42 | 101. | 105. | 165. | 180. | 194. |

Above manhours are for total valve installation, including loadout, handle, haul, flange welds, bolt-ups and valve end weld where required.

See preceding table for suggested valve installation labor crew and equipment spread tables which should be adjusted to fit the individual project needs.

# CLEANING AND TESTING PIPELINES

## General Notes

Consideration should be given to the following items when applying the time requirements for the following labor crew and equipment spread:

1) All pressure testing should be conducted with due regard to the public.
2) Installation of fill line and filling.
3) Sectionalized testing may be required depending on limitation of water supply or to limit the degree of public exposure.
4) Pretesting major road crossings, railroad crossings and prefabricated sections prior to installation may be necessary or desirable.
5) Dewatering of lines with the use of scrapers or spheres.
6) Should defects be found, repair or replace same.
7) Provide radio communication equipment to control and co-ordinate all phases of testing.

# CLEANING AND TESTING PIPELINES

**Labor Crew**

| Personnel Description | Number of Men |
|---|---|
| Foreman | 1 |
| Operator | 1 |
| Oiler | 1 |
| Welder | 6 |
| Welder Helper | 6 |
| Pipefitter | 6 |
| Pipefitter Helper | 6 |
| Swamper | 1 |
| Laborer | 6 |
| Truck Driver | 2 |
| Total Crew | 36 |

The above labor crew is ample for pig cleaning and hydrostatic testing of pipelines.

Length of time required to pig and test line will be determined by length of time required to set up pig launcher and receiving and length and size of pipeline to be tested. This must be estimated for the Individual Project.

# CLEANING AND TESTING PIPELINES

## Equipment Spread

| Equipment Description | Unit | Number of Units |
|---|---|---|
| Light Crane | Each | 1 |
| Pig Launcher | Each | 1 |
| Pig Trap | Each | 1 |
| Pigs | Each | 4 |
| Misc. Pipe, Valves and Fittings | Lot | 1 |
| Anchor Forgings and Tie Downs | Lot | 1 |
| Generator 75 KW | Each | 2 |
| Welding Machine 300 Amp. | Each | 6 |
| Air Compressor 165 CFM | Each | 1 |
| Water Pump—6″ | Each | 2 |
| Test Pump | Each | 2 |
| Miscellaneous Gauges | Lot | 1 |
| Holiday Detector | Each | 2 |
| Truck-Flat Bed | Each | 1 |
| Truck-Pick-Up | Each | 1 |
| Carry All or Bus | Each | 1 |
| Radios (Walkie Talkie) | Each | 2 |

The above equipment spread is ample for pig cleaning and hydrostatic testing of pipelines.

Time requirements, for the above equipment, to pig and test line will be governed by length of time required to set up pig launcher and receiver and length and size of pipeline to be tested. This must be estimated for this individual project.

# BACKFILLING

## General Notes

In applying time requirements to the following labor crew and equipment spread, the following items should be given consideration:

1) Keep backfill as close as possible to lowering-in operation.
2) Hand backfill to six (6) inches above top of pipe.
3) Rock, if available, usually can be used to backfill from six (6) inches above pipe to top of trench.
4) Do not use foreign substances or refuse for backfill material.
5) Place good earth in cultivated areas or where required.
6) Protect backfill from washing.
7) Place riprap when required and where necessary.
8) Cover all bends on the same day they are lowered into trench.

# BACKFILLING

## Labor Crew

| Personnel Description | MEN EACH FOR PIPE SIZES | | |
|---|---|---|---|
| | 4″ to 14″ | 16″ to 24″ | 30″ to 42″ |
| Foreman | 1 | 1 | 1 |
| Operator | 2 | 3 | 3 |
| Oiler | 1 | 1 | 1 |
| Swamper | 2 | 3 | 3 |
| Laborer | 6 | 8 | 10 |
| Total Crew | 12 | 16 | 18 |

## Equipment Spread

| Equipment Description | NUMBER OF UNITS FOR PIPE SIZES | | |
|---|---|---|---|
| | 4″ to 14″ | 16″ to 24″ | 30″ to 42″ |
| Tractor—Front End Loader | 1 | 2 | 2 |
| Dozer D6 or D7 | 1 | 1 | 1 |
| Truck—Pick-up | 1 | 1 | 1 |

Above labor crew and equipment spread is ample for backfilling of trench after pipe is installed assuming pipelay is at the rate of speed as stipulated in the pipelay table, page 4.

If joints are left exposed until after test additional spread time will be required to return and backfill joints.

If additional fill material is required, add hauling labor and equipment to above spreads.

# CLEAN UP OPERATION

## General Notes

In estimating time requirements for clean up operations, consideration should be given to the following items:

1) General clean up of the entire right-of-way.
2) Follow up lowering-in and backfill crews as close as possible.
3) Remove any rubbish, brush, trees, stumps, paint drums and loose rocks.
4) Restore any river or creek banks to their original contour.
5) Where erosion is possible, install riprap or sand bags.
6) Repair damage to or replace fences, terraces, roadways, bridges, culverts, cattle guards, etc.
7) Install pipeline right-of-way markers.

# CLEAN UP OPERATION

**Labor Crew**

| Personnel Description | MEN EACH FOR RIGHT-OF-WAY | | |
|---|---|---|---|
| | 50 L.F. Width | 80 L.F. Width | 100 L.F. Width |
| Foreman | 1 | 2 | 3 |
| Operator | 4 | 7 | 8 |
| Oiler | 1 | 2 | 3 |
| Swamper | 3 | 6 | 6 |
| Truck Driver | 1 | 3 | 3 |
| Laborer | 10 | 16 | 20 |
| Total Crew | 20 | 36 | 43 |

Above total crew is ample for cleaning up operations of construction debris for one (1) mile of right-of-way per ten (10) hour day for the width as outlined above.

Above manpower does not include hauling or burning of brush and trees. See clear and grade right-of-way table for this operation, page 11.

**Equipment Spread**

| Equipment Description | NUMBER OF UNITS FOR RIGHT-OF-WAY | | |
|---|---|---|---|
| | 50 L.F. Width | 80 L.F. Width | 100 L.F. Width |
| Dozer D7 or D8 | 1 | 4 | 5 |
| Dozer D6 | 2 | 2 | 2 |
| Front End Loader | 1 | 1 | 1 |
| Truck—2 Ton | 1 | 3 | 3 |
| Truck—Dump | 1 | 1 | 1 |
| Truck—Pick-up | 1 | 2 | 2 |

Above equipment spread is ample for clean up operations of construction debris for one (1) mile of right-of-way per ten (10) hour day for the width as outlined above.

Above equipment spread does not include equipment for hauling of brush and trees. See clear and grade right-of-way table for this operation, page 11.

# UTILITY OPERATIONS

## General Notes

Utility operations can vary greatly, depending on the obstructions that may lie in the way of the pipeline. Therefore, adjustment to the following labor crew and equipment spread may be required for:

1) Boring under highways, roads, railroads, etc.
2) Placing conductor piping or casings.
3) Break out and remove pavements and repair and replace damage to same.
4) Install and maintain safety precautions as required such as lanterns, barricades, warning signs, etc.

# UTILITY OPERATIONS

## Labor Crew

| Personnel Description | MEN EACH FOR PIPE SIZES | | |
|---|---|---|---|
| | 4″ to 14″ | 16″ to 24″ | 30″ to 42″ |
| Foreman | 1 | 1 | 1 |
| Mechanic | 1 | 1 | 1 |
| Operator | 7 | 8 | 8 |
| Oiler | 1 | 1 | 1 |
| Air Tool Operator | 2 | 2 | 2 |
| Swamper | 4 | 5 | 5 |
| Welder | 2 | 3 | 4 |
| Welder Helper | 2 | 3 | 4 |
| Truck Driver | 2 | 3 | 4 |
| Laborer | 4 | 6 | 8 |
| Total Crew | 26 | 33 | 38 |

Above labor crew is average for installation of casements and miscellaneous borings and other utility type operations which may be involved for several hundred miles of cross-country pipeline of the sizes as listed above.

Number and type of utility operations for the individual project will govern the size of crew required for each project. Consideration should be given to the individual project and adjustments made to the labor crew accordingly.

# UTILITY OPERATIONS

## Equipment Spread

| Equipment Description | NUMBER OF UNITS FOR PIPE SIZES | | |
|---|---|---|---|
| | 4″ to 14″ | 16″ to 24″ | 30″ to 42″ |
| Dragline | 1 | 1 | 1 |
| Side Boom | 2 | 3 | 3 |
| Dozer | 1 | 1 | 1 |
| Welding Machine | 2 | 3 | 4 |
| Boring Machine | 1 | 1 | 1 |
| Cut and Bevel Outfit | 1 | 1 | 1 |
| Asphalt Kettle | 1 | 1 | 1 |
| Air Compressor | 1 | 1 | 1 |
| Truck—6 × 6 | 1 | 1 | 1 |
| Truck—2 Ton | 1 | 2 | 3 |
| Truck—Pick-up | 1 | 1 | 1 |

Above equipment spread is average for installation of casements and miscellaneous borings and other utility type operations as may be involved for several hundred miles of cross-country pipeline of the sizes as listed above.

Number and type of utility operations for the individual project will govern the size and amount of equipment that will be required for each project. Consideration should be given to the individual project and adjustments made to the equipment spread accordingly.

Boring machine is to include the necessary augers, auger bushings, connectors and adapters and cutting heads.

# Section Two

# MARSHLAND PIPELINES

The following labor crews and equipment spreads are intended to suffice only for those direct erection operations that may be necessary for the installation of pipelines in marshland or swampy areas.

Labor forces and equipment for operations such as bending, cleaning, priming, coating and wrapping, sandblasting and painting, etc., can be obtained from the tables included under Section One of this manual and added to the spreads in this section if necessary.

The following labor crews are for a single twelve (12) hour shift and the equipment listings are for a single spread. More than one crew and spread may be required, depending on project scope and schedule.

All spreads are to be equipped with short-wave radio communication devices capable of maintaining contact with all required office and storage facilities as well as with the individual spreads.

Normal weather conditions are usually considered and included in the base price. Downtime for all marine equipment due to severe weather, stand-by due to no fault of the contractor, and major breakdown time are usually quoted to the client as a reduced non-working day rate or are covered in the price by an allowance of an estimated number of days at the reduced non-working day rate.

To determine the daily cost of a spread for a non-working day simply reduce the working day spread rate by the cost of the productive operating fuel, oil and grease, the consumable supplies, and any indirect cost and fee that you feel is justifiable.

# AVERAGE PIPELAY TABLE

**Linear Feet of Pipelay per Twelve (12) Hour Day in Trench in Non-Obstructive Marshland**

| | PIPE WALL THICKNESS IN INCHES | | | | |
|---|---|---|---|---|---|
| Nominal Pipe Size | 0.000 Through 0.250 | 0.251 Through 0.375 | 0.376 Through 0.500 | 0.501 Through 0.625 | 0.626 Through 0.750 |
| 4 | 3,780 | 3,675 | 3,600 | — | — |
| 6 | 3,360 | 3,290 | 3,230 | — | — |
| 8 | 3,200 | 3,120 | 3,050 | 2,990 | — |
| 10 | 3,120 | 3,070 | 3,000 | 2,940 | — |
| 12 | 3,010 | 2,940 | 2,880 | 2,820 | 2,770 |
| 14 | 2,940 | 2,880 | 2,850 | 2,800 | 2,750 |
| 16 | 2,800 | 2,750 | 2,690 | 2,640 | 2,580 |
| 18 | 2,730 | 2,680 | 2,620 | 2,570 | 2,520 |
| 20 | 2,590 | 2,540 | 2,490 | 2,440 | 2,390 |
| 24 | — | 2,310 | 2,260 | 2,220 | 2,170 |
| 30 | — | 2,100 | 2,060 | 2,020 | 1,970 |
| 36 | — | — | 1,890 | 1,860 | 1,820 |
| 42 | — | — | 1,680 | 1,650 | 1,610 |

The construction labor and equipment spreads in this section should be capable of performing all operations that may be involved in the installation of the above quantities of pipe in a twelve (12) hour work day, in unobstructed marshland.

Productivity will vary where obstructions such as heavy grass, tree stumps, etc., are encountered.

See following respective tables and explanations covering productivity factors to be applied for variation in obstructions.

See pages xi-xiv for explanation covering productivity factors to be applied for labor.

# PRODUCTIVITY AND PERCENTAGE EFFICIENCY FACTORS

To correctly apply the time frames that have been included in the preceding tables consideration must be given to the effects of labor productivity and to obstructions that may lie within the path or right-of-way of the pipeline.

Since the effect of these items has a decisive bearing on the actual linear feet of pipelay one might expect to obtain per day, we have included the following percentage efficiency tables covering these items and an example of their application.

Productivity factors to be applied to linear feet of pipelay per day for labor are as follows:

LABOR PRODUCTIVITY FACTORS

| Type of Productivity | Percentage Range |
|---|---|
| Excellent | 0.901 through 1.000 |
| Very good | 0.801 through 0.900 |
| Average | 0.601 through 0.800 |
| Low | 0.401 through 0.600 |
| Very low | 0.001 through 0.400 |

The following production elements should be given consideration for the application of a labor productivity percentage.

1. General Economy
2. Project Supervision
3. Labor Conditions
4. Job Conditions
5. Equipment
6. Weather

For a full description of above elements and an example of obtaining labor productivity percentage see pages xi-xiv.

Obstruction productivity factors to be applied to linear feet of pipelay per day are as follows:

OBSTRUCTION PRODUCTIVITY FACTORS

| Description of Marshland | Percent of Obstructions | Productivity Factor |
|---|---|---|
| Grass | 0 | 1.000 |
| Grass—Few Small Stumps | 0 to 25 | 0.875 |
| Grass—Few Medium Stumps | 25 to 50 | 0.750 |
| Grass—Few Large Stumps | 50 to 75 | 0.600 |
| Grass—Many Large Stumps | 75 to 100 | 0.450 |

*Grass:* No tree stumps or roots—grass only.

*Grass—Few Small Stumps:* Grass with a few minor small stumps. Easy to remove.

*Grass—Few Medium Stumps:* Grass with a few medium-size stumps. Not too easy to remove.

*Grass—Few Large Stumps:* Grass with a few large stumps. Hard to remove.

*Grass—Many Large Stumps:* Grass with many large stumps. Very hard to remove.

# EXAMPLE—APPLICATION OF FACTORS

Due to the existence of such a wide degree of variability between labor and obstruction factors, we offer the following example of a method that may be used to determine the actual linear feet of pipelay that should be installed in a day with certain labor and obstruction variables.

For our example we assume that we have one hundred (100) miles of 30-inch by ⅜-inch wall pipe to install under the following conditions and that we will be working a twelve (12) hour day.

From the preceding average pipelay table, we determine that twenty one hundred (2100) linear feet of 30-inch line can be installed in unobstructed marshland in a twelve (12) hour day.

From the labor analyzation table (see pages xi-xiv), it has been determined that our labor productivity is equal to sixty-five (65) percent.

# EXAMPLE—APPLICATION OF FACTORS (continued)

Let us assume that the total pipeline is not to be installed in total unobstructed marshland and that the following obstructions exist:

OBSTRUCTION FACTORS

| Description | Obstruction Factor | Miles of Pipe | Use Factor |
|---|---|---|---|
| Grass | 1.000 | 50.0 | 50.0 |
| Grass—Few Small Stumps | 0.875 | 20.0 | 17.5 |
| Grass—Few Medium Stumps | 0.750 | 20.0 | 15.0 |
| Grass—Few Large Stumps | 0.600 | 2.0 | 1.2 |
| Grass—Many Large Stumps | 0.450 | 8.0 | 3.6 |
| Total | — | 100.0 | 87.3 |

$$\text{Obstruction Factor} = \frac{87.3}{100} = 0.873$$

Average rate of pipelay per twelve (12) hour day based on the above assumption is equal to:

Linear Feet Per Day × Labor Factor × Obstruction Factor

or

2100 L.F. × 0.650 × 0.873 = 1192 L.F. Per Day

The direct installation cost, in dollar value per linear foot of pipe, can be obtained by the following formula:

$$\text{Direct Cost Per Foot} = \frac{\text{Cost Per Day of Spreads}}{\text{Linear Feet Per Day of Lay}}$$

# ALIGNING AND WELDING OPERATIONS

## General Notes

In keeping with the average pipelay per day table (page 46), consideration should be given to the following items for all aligning and welding operations:

1) The labor crews and equipment spreads are based on welds being made by the manual shielded metal-arc method.
2) Welding machines should be of the size and type suitable for the work.
3) Welding machines are to be operated within the amparage and voltages ranges recommended for each size and type of electrode.
4) Remove all foreign matter from the beveled ends before welding. Clean with hand or power tools.
5) Pipe ends must have proper bevels for the work intended.
6) The alignment of abutting pipe ends should be such as to minimize the offset between pipe surface. Hammering of pipe to obtain proper line-up should be held to a minimum.
7) Internal line-up clamps should be used where possible. Use external line-up clamps when it is impractical or impossible to use an internal line-up clamp.
8) Use minimum of root bead, filler bead and finish bead.
9) Brush and clean the finished weld.
10) Only qualified welders should be used.
11) Where required apply heat treatment-preheating and postheating.
12) Apply radiographic inspection of welds where required.

# ALIGNING AND WELDING OPERATIONS

**Labor Crew**

| Personnel Description | Number of Men Required |
|---|---|
| Barge Foreman | 1 |
| Welder Foreman | 1 |
| Welders | 6 |
| Welder Helper (Apprentice) | 6 |
| Spacer | 1 |
| Stabber | 1 |
| Buffer | 6 |
| Mechanic | 1 |
| Operator—Anchors | 3 |
| Operator—Equipment | 2 |
| Operator—Equipment Tender | 3 |
| Oiler | 2 |
| Kettle Fireman | 1 |
| Swamper | 2 |
| Labor Foreman | 1 |
| Laborer | 12 |
| Truck Driver | 2 |
| X-Ray Technician | 1 |
| Total Crew | 52 |

The above labor crew is ample for the complete welding and installation operations required for pipelines in marshlands all in accordance with the average pipelay table and obstruction and labor productivity factors pages 46-51.

The above manpower can be adjusted upward or downward depending on amount of pipe to be installed, project schedule and installation specifications.

In all cases ample manpower must be allowed for operation of all equipment that may be listed in the equipment spread.

# ALIGNING AND WELDING OPERATIONS

## Equipment Spread

| Equipment Description | Number of Units Required |
|---|---|
| Tugs (Fully Manned)—400 hp± | 3 |
| Lay Barge | 1 |
| Material Barge—120′ × 30′ ± | 1 |
| Cargo Barge—110′ × 30′ ± | 1 |
| Crew Boat—40′ ± | 2 |
| Skiff With Motor | 1 |
| Crane—50 Ton | 1 |
| Clam Bucket—¾ Cy. | 1 |
| Drag Bucket—1½ Cy. | 1 |
| Side Boom—572 | 1 |
| Welding Machine—300 AMP. | 8 |
| Generator—6 KW. | 1 |
| Water Pump—2″ | 3 |
| Water Pump—4″ | 1 |
| Air Compressor—165 CFM. | 1 |
| Line Up Clamps | 2 |
| Launch Shoes | 6 |
| Hydraulic Shoe | 1 |
| Tar Kettle—10 Barrel | 1 |
| Cement Mixer—6 S | 1 |
| Holiday Detectors | 3 |
| Pick-up Truck | 4 |
| Mechanics Truck | 1 |
| X-Ray Equipment—Set Up | 1 |

The above equipment spread is ample for the complete welding and installation operations required for pipelines in marshlands all in accordance with the average pipelay table as may be amended by the obstruction and labor productivity factors, pages 46-51.

The above equipment can be adjusted upward or downward depending on the amount of pipe to be installed, project schedule and installation specification conditions.

# ALIGNING AND WELDING OPERATIONS— LAND SHOVE METHOD

**Labor Crew**

| Personnel Description | Number of Men Required |
|---|---|
| Crew Foreman | 1 |
| Welder Foreman | 1 |
| Welder | 5 |
| Welder Helper (Apprentice) | 5 |
| Spacer | 1 |
| Stabber | 1 |
| Buffer | 5 |
| Pot Fireman | 1 |
| Operator—Equipment | 2 |
| Operator—Equipment Tender | 1 |
| Oiler | 1 |
| Swamper | 1 |
| Laborer | 12 |
| Truck Driver | 2 |
| X-Ray Technician | 1 |
| Total Crew | 40 |

The above labor crew is ample for pushing pipe and all other operations involved in the installation of pipe by this method in marshlands or small river and stream crossings.

The above manpower can be adjusted upward or downward depending on length of run of pipe to be installed by this method, project schedule and installation specifications.

In all cases ample manpower must be allowed for operating all selected equipment for this type of operation.

# ALIGNING AND WELDING OPERATIONS—LAND SHOVE METHOD

### Equipment Spread

| Equipment Description | Number of Units Required |
|---|---|
| Side Boom 572 | 1 |
| Crane With ¾ Cy. Drag Bucket | 1 |
| Welding Machine—300 AMP. | 6 |
| Air Compressor—165 CFM. | 1 |
| Generator—6 KW. | 1 |
| Hoist | 1 |
| Tar Kettle—3 Barrel | 1 |
| Cement Mixer—6 S | 1 |
| Jet Pump—6″ | 1 |
| Water Pump—2″ | 1 |
| Launch Shoes | 6 |
| Holiday Detector | 2 |
| Skiff With Motor | 1 |
| Pick-up Truck | 2 |
| Flatbed Truck—2½ Ton | 1 |
| Bus | 1 |
| X-Ray Equipment Set Up | 1 |

The above equipment spread is ample for pushing pipe and all other operations involved in the installation of pipe by this method in marshlands or small river and stream crossings.

The above equipment can be adjusted upward or downward depending on length of run of pipe to be installed by this method, project schedule and installation specification conditions.

# ALIGNING AND WELDING OPERATIONS—RIVER CROSSING TIE-IN

**Labor Crew**

| Personnel Description | Number of Men Required |
|---|---|
| Barge Foreman | 1 |
| Welder | 2 |
| Welder Helper (Apprentice) | 2 |
| Operator—Anchor | 2 |
| Operator—Equipment | 1 |
| Operator—Equipment Tender | 1 |
| Oiler | 1 |
| Spacer | 1 |
| Stabber | 1 |
| Buffer | 2 |
| Pot Fireman | 1 |
| Laborer | 6 |
| Truck Driver | 1 |
| Total Crew | 22 |

The above labor crew is ample for all operations that may be involved in river or lake crossing installation of pipelines.

Manpower can be adjusted upward or downward depending on length of run and size of pipe to be installed by this method, project schedule and installation specification conditions.

In all cases ample manpower must be allowed for operating all selected equipment for this type of operation.

# ALIGNING AND WELDING OPERATIONS—RIVER CROSSING TIE-IN

**Equipment Spread**

| Equipment Description | Number of Units Required |
|---|---|
| Tug—400 H.P. (Fully Manned) | 1 |
| Utility Barge | 2 |
| Cargo Barge—110′ × 30′ | 1 |
| Crew Boat—40′ | 1 |
| Skiff With Motor | 2 |
| Crane—Dragline—30 Ton | 1 |
| Clam Bucket—¾ Cy. | 1 |
| Drag Bucket—1½ Cy. | 1 |
| Air Compressor—165 CFM. | 1 |
| Welding Machine—300 AMP. | 3 |
| Generator—17 KW. | 1 |
| Generator—6 KW. | 1 |
| Jet Pump—6″ | 1 |
| Water Pump—2″ | 2 |
| Tar Kettle—3 Barrel | 1 |
| Cement Mixer—6 S | 1 |
| Holiday Detector | 2 |
| Hoist—G140 | 1 |
| Pick-up Truck | 1 |

The above equipment spread is ample for all operations that may be involved in river or lake crossing installation of pipelines.

Equipment can be adjusted upward or downward depending on length or run and size of pipe to be installed by this method, project schedule and installation specification conditions.

# TUG

## Labor Crew

| Personnel Description | Number of Men Required |
|---|---|
| Captain | 1 |
| Mate | 1 |
| Chief Engineer | 1 |
| First Engineer | 1 |
| Seaman | 2 |
| Oiler | 1 |
| Cook | 1 |
| Total Crew | 8 |

## Equipment Spread

| Equipment Description | Number of Units Required |
|---|---|
| Tug | 1 |

The above labor crew is standard for operating an ocean certified tug.

For close-to-shore or inland waters, work this crew can be reduced by deletion of the first engineer and cook if work is to be performed on a one-shift basis.

When tugs are rented from outside sources, the rental rate should include the full contingent of personnel and be completely rigged for the type of operation and work it is to perform.

# HOE DITCH OPERATION

### Labor Crew

| Personnel Description | Number of Men Required |
|---|---|
| Ditch Foreman | 1 |
| Mechanic | 1 |
| Operator—Backhoe | 4 |
| Operator—Skiff | 4 |
| Laborer | 4 |
| Truck Driver | 3 |
| Total Crew | 17 |

### Equipment Spread

| Equipment Description | Number of Units Required |
|---|---|
| Tug—165 HP. (Fully Manned) | 1 |
| Material Barge | 1 |
| Skiff With Motor | 4 |
| Crane—Backhoe | 4 |
| Mechanics Truck | 1 |
| Pick-up Truck | 1 |
| Carry All or Van | 3 |

The above labor crew and equipment spread is ample for ditching operations in mud, mud and sand and light clay type soil materials as are usually found in marshland areas.

Length of time required for the above crew and spread will be determined by the type of soil, width and depth of ditch and length of run.

Labor crew and equipment spread can be adjusted upward or downward depending on the type of soils and project schedule.

Do not open ditch too far in advance of pipelay crew.

Labor crew and equipment spread similar to the above can be used for backfilling.

# VALVE INSTALLATION

## Labor Crew

| Personnel Description | Number of Men Required |
|---|---|
| Foreman | 1 |
| Pipefitter—Welder | 6 |
| Pipefitter—Helper | 6 |
| Operator | 1 |
| Oiler | 1 |
| Laborer | 4 |
| Truck Driver | 1 |
| Total Crew | 20 |

## Equipment Spread

| Equipment Description | Number of Units Required |
|---|---|
| Light Crane | 1 |
| Welding Machine—300 AMP. | 1 |
| Generator | 1 |
| Flatbed Truck—2 Ton | 1 |
| Pick-up Truck | 1 |

The above labor crew and equipment spread should be ample for the installation of valves in pipelines.

Number of valves to be installed, valve size, type and location will govern the actual size of crew and equipment spread required.

See page 62 for average hourly time requirements for valve installation.

# VALVE INSTALLATION

**Average Hourly Time Requirements**

| Nominal Pipe Size | FLANGED AND BOLTED VALVES | | | | |
|---|---|---|---|---|---|
| | SERVICE PRESSURE RATING | | | | |
| | 150 Lb. | 300-400 Lb. | 600-900 Lb. | 1500 Lb. | 2500 Lb. |
| 8 | 17. | 19. | 25. | 35. | 40. |
| 10 | 19. | 24. | 34. | 44. | 50. |
| 12 | 27. | 29. | 42. | 55. | 63. |
| 14 | 31. | 35. | 49. | 63. | 73. |
| 16 | 36. | 40. | 56. | 72. | 85. |
| 18 | 41. | 46. | 64. | 85. | 98. |
| 20 | 48. | 54. | 73. | 108. | 110. |
| 22 | 53. | 60. | 83. | 111. | 114. |
| 24 | 58. | 65. | 92. | 121. | 124. |
| 26 | 65. | 71. | 100. | 123. | — |
| 30 | 79. | 86. | 119. | 144. | — |
| 36 | 100. | 108. | 148. | 177. | — |
| 42 | 121. | 130. | 176. | 211. | — |
| | WELD END VALVES | | | | |
| 8 | 8. | 9. | 10. | 11. | 13. |
| 10 | 10. | 10. | 13. | 15. | 17. |
| 12 | 13. | 13. | 17. | 19. | 23. |
| 14 | 15. | 16. | 21. | 24. | 31. |
| 16 | 19. | 20. | 26. | 30. | 38. |
| 18 | 24. | 25. | 33. | 38. | 49. |
| 20 | 27. | 28. | 40. | 46. | 57. |
| 22 | 31. | 32. | 47. | 53. | 65. |
| 24 | 35. | 37. | 54. | 61. | 72. |
| 26 | 40. | 42. | 69. | 75. | 93. |
| 30 | 49. | 51. | 82. | 89. | 109. |
| 36 | 69. | 72. | 113. | 125. | 141. |
| 42 | 101. | 105. | 165. | 180. | 194. |

The above manhours are for total valve installation including loadout, handle, haul, flange welds, bolt-ups and valve end weld where required.

See preceding table for suggested valve installation labor crew and equipment spread tables which should be adjusted to fit the individual project needs.

# CLEANING AND TESTING PIPELINES

## General Notes

Consideration should be given to the following items when applying the time requirements for the following labor crew and equipment spread:

1) All pressure testing should be conducted with due regard for the public.
2) Installation of fill lines and filling operation.
3) Sectionalized testing may be required depending on limitation of water supply.
4) Pretesting of water crossings or prefabricated sections may be required or desired.
5) Dewatering of lines with the use of scrapers or spheres.
6) Should defects be found repair or replace same.
7) Provide radio communication equipment to control and co-ordinate all phases of testing.

# CLEANING AND TESTING PIPELINES

**Labor Crew**

| Personnel Description | Number of Men Required |
|---|---|
| Foreman | 1 |
| Operator | 1 |
| Oiler | 1 |
| Welder | 6 |
| Welder Helper | 6 |
| Pipefitter | 6 |
| Pipefitter Helper | 6 |
| Swamper | 1 |
| Laborer | 6 |
| Truck Driver | 2 |
| Total Crew | 36 |

The above labor crew is ample for pig cleaning and hydrostatic testing of pipelines.

Length of time required to pig and test line will be determined by length of time required to set up pig launcher and receiver and length and size of pipeline to be tested. This must be estimated for the individual project.

# CLEANING AND TESTING PIPELINES

## Equipment Spread

| Equipment Description | Unit | Number of Units Required |
|---|---|---|
| Light Crane | Each | 1 |
| Pig Launcher | Each | 1 |
| Pig Trap | Each | 1 |
| Pigs | Each | 4 |
| Misc. Pipe, Valves and Fittings | Lot | 1 |
| Anchor Forgings and Tie Downs | Lot | 1 |
| Generator 75 KW. | Each | 2 |
| Welding Machine—300 AMP. | Each | 6 |
| Air Compressor—165 CFM. | Each | 1 |
| Water Pump—6″ | Each | 2 |
| Test Pump | Each | 2 |
| Miscellaneous Gauges | Lot | 1 |
| Holiday Detector | Each | 2 |
| Truck—Flatbed | Each | 1 |
| Truck—Pick-up | Each | 1 |
| Carry All or Bus | Each | 1 |
| Radios (Walkie Talkie) | Each | 2 |

The above equipment spread is ample for pig cleaning, and hydrostatic testing of pipelines.

Time requirements for the above equipment to pig and test line will be governed by length of time required to set up pig launcher and receiver and length and size of pipeline to be tested. This must be estimated for the individual project.

# Section Three

# ONSHORE AND OFFSHORE SURF-ZONE PIPELINES

This section provides labor crew and equipment spread tables for the installation of onshore pipelines running a short distance from storage or blending facilities and continuing through the offshore surf-zone area only.

No consideration has been given to the installation of storage and blending facilities or pumping stations. Construction methods, labor, materials and equipment for these facilities vary greatly from those required for pipeline installation and therefore should be evaluated for the particular project and estimated accordingly.

Landlay pipeline crews normally work a single shift of ten (10) hours per day. Because of the high cost of marine equipment, the vagaries of the weather, etc., offshore pipelay crews usually work two (2) shifts of twelve (12) hours each for a twenty-four (24) hour day total.

The manpower listed in the tables is for one crew, working a single shift only. Should a land crew be assisted by or support an offshore spread, the land crews and equipment spreads should be adjusted in cost to equal the time frame of the offshore spread.

It is assumed that at least one pick-up truck in each spread will be equipped with short-wave radio equipment capable of communicating with all required office, storage and individual spread facilities and that each spread will be equipped with walkie-talkie radios for communicating between the various spreads.

# AVERAGE PIPELAY TABLE—ON LAND, NEAR SHORE, UNDERGROUND

**Linear Feet of Pipelay Per Ten (10) Hour Day**
**In Trench in Level Rock-Free Terrain**

| | PIPE WALL THICKNESS IN INCHES | | | | | |
|---|---|---|---|---|---|---|
| Nominal Pipe Size | 0.000 through 0.250 | 0.251 through 0.375 | 0.376 through 0.500 | 0.501 through 0.625 | 0.626 through 0.750 | 0.756 through 1.000 |
| 4 | 4,320 | 4,200 | 4,120 | — | — | — |
| 6 | 3,840 | 3,760 | 3,690 | — | — | — |
| 8 | 3,650 | 3,560 | 3,490 | 3,420 | — | — |
| 10 | 3,560 | 3,510 | 3,430 | 3,360 | — | — |
| 12 | 3,440 | 3,360 | 3,290 | 3,230 | 3,160 | — |
| 14 | 3,360 | 3,290 | 3,270 | 3,200 | 3,140 | — |
| 16 | 3,200 | 3,130 | 3,070 | 3,010 | 2,950 | 2,890 |
| 18 | 3,120 | 3,060 | 3,000 | 2,940 | 2,880 | 2,820 |
| 20 | 2,960 | 2,900 | 2,840 | 2,790 | 2,730 | 2,670 |
| 24 | — | 2,640 | 2,590 | 2,540 | 2,490 | 2,430 |
| 30 | — | 2,400 | 2,350 | 2,300 | 2,260 | 2,210 |
| 36 | — | — | 2,160 | 2,120 | 2,080 | 2,040 |
| 42 | — | — | 1,920 | 1,880 | 1,840 | 1,800 |

The construction labor and equipment spreads that follow in this section should provide for all operations in the installation of the above quantities of pipe in a ten (10) hour work day, in level rock-free ground, from shoreline to storage or process facilities located a maximum of two (2) miles inland from shoreline.

Productivity will vary where different types of terrain and/or rock are encountered.

See following respective tables and explanations covering productivity factors to be applied for variation in terrain and rock.

See pages xi-xiv for explanation covering productivity factors to be applied for labor.

Above footage is based on installing double random joints of pipe.

# AVERAGE PIPELAY TABLE— ON LAND, NEAR SHORE, ABOVE GROUND ON SUPPORTS

**Linear Feet of Pipelay Per Ten (10) Hour Day**
**On Supports to Waist-High, On Level Ground**

| | PIPE WALL THICKNESS IN INCHES | | | | | |
|---|---|---|---|---|---|---|
| Nominal Pipe Size | 0.000 through 0.250 | 0.251 through 0.375 | 0.376 through 0.500 | 0.501 through 0.625 | 0.626 through 0.750 | 0.751 through 1.000 |
| 4 | 4,530 | 4,440 | 4,350 | — | — | — |
| 6 | 4,030 | 3,952 | 3,870 | — | — | — |
| 8 | 3,830 | 3,750 | 3,680 | 3,600 | — | — |
| 10 | 3,760 | 3,680 | 3,610 | 3,540 | — | — |
| 12 | 3,520 | 3,450 | 3,400 | 3,330 | 3,270 | — |
| 14 | 3,480 | 3,430 | 3,360 | 3,300 | 3,240 | — |
| 16 | 3,220 | 3,150 | 3,100 | 3,030 | 2,980 | 2,920 |
| 18 | 3,120 | 3,060 | 3,000 | 2,940 | 2,880 | 2,830 |
| 20 | 3,020 | 2,960 | 2,900 | 2,840 | 2,790 | 2,730 |
| 24 | — | 2,690 | 2,640 | 2,590 | 2,530 | 2,480 |
| 30 | — | 2,450 | 2,400 | 2,350 | 2,310 | 2,270 |
| 36 | — | — | 2,180 | 2,140 | 2,100 | 2,060 |
| 42 | — | — | 1,940 | 1,900 | 1,860 | 1,830 |

The construction labor and equipment spreads that follow in this section should provide for all operations involved in the installation of the above quantities of pipe in a ten (10) hour work day, on waist-high supports on level ground, from shoreline to storage or process facilities located a maximum of two (2) miles inland from shoreline.

Productivity will vary depending on contour of terrain and accessibility.

See following respective tables and explanation covering productivity factors to be applied for variation in terrain.

See pages xi-xiv for explanation covering productivity factors to be applied for labor.

Above footage is based on installing double random joints of pipe.

# AVERAGE PIPELAY TABLE— OFFSHORE SUBMERGED SURF-ZONE PIPE LAND FABRICATE-BARGE PULL

**Linear Feet of Concrete Weight Coated Pipe Per Twenty-Four (24) Hour Day In Pre-Excavated Trench or On Bottom**

| Nominal Pipe Size | PIPE WALL THICKNESS IN INCHES | | | | | |
|---|---|---|---|---|---|---|
| | 0.250 | 0.251 through 0.375 | 0.376 through 0.500 | 0.501 through 0.625 | 0.626 through 0.750 | 0.756 through 1.000 |
| 4 | 3,810 | 3,700 | 3,560 | — | — | — |
| 6 | 3,370 | 3,300 | 3,180 | — | — | — |
| 8 | 3,210 | 3,120 | 3,060 | 3,000 | — | — |
| 10 | 3,150 | 3,090 | 3,000 | 2,940 | — | — |
| 12 | 3,000 | 2,970 | 2,880 | 2,820 | 2,740 | — |
| 14 | 2,940 | 2,910 | 2,850 | 2,760 | 2,700 | — |
| 16 | 2,800 | 2,770 | 2,680 | 2,600 | 2,550 | 2,500 |
| 18 | 2,740 | 2,640 | 2,620 | 2,550 | 2,500 | 2,450 |
| 20 | 2,610 | 2,590 | 2,490 | 2,420 | 2,370 | 2,320 |
| 24 | — | 2,460 | 2,270 | 2,200 | 2,160 | 2,120 |
| 30 | — | 2,240 | 2,070 | 2,000 | 1,960 | 1,920 |
| 36 | — | — | 1,900 | 1,840 | 1,800 | 1,760 |
| 42 | — | — | 1,690 | 1,640 | 1,610 | 1,580 |

The related construction labor and equipment spreads that follow should provide for all operations, excluding trench blasting that may be involved in the installation of the above quantities of pipe in the offshore surf-zone area, a maximum distance of three thousand (3000) linear feet, working a double twelve (12) hour shift or twenty-four (24) hour day total.

The above linear feet of pipelay is based on assembling concrete weight coated pipe on land and pulling with winch and cables from an anchored barge offshore specifically rigged with proper equipment for this type operation.

Above footage is based on installing double random joints of pipe.

# AVERAGE PIPELAY TABLE— OFFSHORE SUBMERGED SURF-ZONE PIPE LAY BARGE FABRICATE-LAND PULL

**Linear Feet of Concrete Weight Coated Pipe Per Twenty-Four (24) Hour Day In Pre-Excavated Trench or On Bottom**

| | PIPE WALL THICKNESS IN INCHES | | | | | |
|---|---|---|---|---|---|---|
| Nominal Pipe Size | 0.250 | 0.251 through 0.375 | 0.376 through 0.500 | 0.501 through 0.625 | 0.626 through 0.750 | 0.751 through 1.000 |
| 4 | 4,950 | 4,810 | 4,630 | — | — | — |
| 6 | 4,380 | 4,290 | 4,130 | — | — | — |
| 8 | 4,180 | 4,060 | 3,980 | 3,900 | — | — |
| 10 | 4,100 | 4,020 | 3,900 | 3,820 | — | — |
| 12 | 3,900 | 3,860 | 3,740 | 3,670 | 3,560 | — |
| 14 | 3,820 | 3,780 | 3,700 | 3,590 | 3,510 | — |
| 16 | 3,640 | 3,600 | 3,480 | 3,380 | 3,320 | 3,250 |
| 18 | 3,560 | 3,430 | 3,400 | 3,320 | 3,250 | 3,190 |
| 20 | 3,400 | 3,370 | 3,240 | 3,150 | 3,080 | 3,020 |
| 24 | — | 3,200 | 2,950 | 2,860 | 2,810 | 2,760 |
| 30 | — | 2,920 | 2,690 | 2,600 | 2,550 | 2,500 |
| 36 | — | — | 2,470 | 2,390 | 2,340 | 2,290 |
| 42 | — | — | 2,200 | 2,130 | 2,090 | 2,050 |

The related construction labor and equipment spreads that follow in this section should provide for all operations, excluding trench blasting, that may be involved in the installation of the above quantities of pipe in the offshore surf-zone area, a maximum distance of three thousand (3000) linear feet, working a double twelve (12) hour shift or twenty-four (24) hour day total.

The above linear feet pipelay is based on assembling concrete weight coated pipe on a pipelay barge and pulling from the beach with pre-anchored winches and cables specifically rigged with the proper equipment for this type operation.

Above footage is based on installing double random joints of pipe.

# PRODUCTIVITY AND PERCENTAGE EFFICIENCY FACTORS

As previously stated, the preceding time frame tables are based on one hundred percent productivity efficiency. Due to many factors this is seldom if ever achieved. In the installation of onshore beach area and offshore surf-zone area pipelay, we have found that the most important items that effect the percentage of efficiency are labor, both on and offshore, terrain and rock conditions onshore and bottom conditions offshore.

The following productivity efficiency factor tables covering labor, onshore terrain and rock, and offshore bottom can be cross-referenced with the preceding pipelay tables to assist the estimator in arriving at the actual average total linear feet per day of pipelay that should be obtained. An example of the application of these factors is included.

In some areas wind, wave, and currents may become serious factors. If this is the case, see Section Four—OCEAN AND SEA PIPELINES.

Productivity factors to be applied to labor for linear feet of onshore and offshore surf-zone pipelay per day are as follows:

LABOR PRODUCTIVITY FACTORS
NEAR SHORE, ON-LAND PIPELAY

| Type of Productivity | Percentage Range |
|---|---|
| Excellent | 0.901 Through 1.000 |
| Very Good | 0.801 Through 0.900 |
| Average | 0.601 Through 0.800 |
| Low | 0.401 Through 0.600 |
| Very Low | 0.001 Through 0.400 |

The following production elements should be given consideration for the application of labor productivity percentage:

1. General Economy
2. Project Supervision
3. Labor Conditions
4. Job Conditions
5. Equipment
6. Weather

For full description of above elements and example of obtaining labor productivity percentage see pages xi-xiv.

Productivity factors for terrain to be applied to linear feet of near-shore, on-land pipelay per day are as follows:

TERRAIN PRODUCTIVITY FACTORS
NEAR-SHORE, ON-LAND PIPELAY

| Description Of Terrain | Productivity Factor |
|---|---|
| Level Sandy Loam | 1.000 |
| Sandy Loam With Sand Dunes | 0.950 |
| Muck and Mud | 0.800 |
| Minor Up and Down Grade Changes | 0.750 |
| Steep Upgrade Change | 0.500 |

*Level Sandy Loam:* Smooth and level with constant minor upgrade.

*Sandy Loam With Sand Dunes:* Constant contour changes with small sand dunes.

*Muck and Mud:* Dirty slush and wet sticky mud.

*Minor Up-and Down-Grade Changes:* Partially level with minor contour changes, slues and valleys.

*Steep Up-Grade Changes:* Constant steep up-grade changes, major hills, slues and valleys.

Productivity factors to be applied for rock to linear feet of near-shore, on-land pipelay per day are as follows:

ROCK PRODUCTIVITY FACTORS
NEAR-SHORE, ON-LAND PIPELAY

| Percent of Rock | Productivity Factor |
|---|---|
| 0 | 1.000 |
| 0 to 25 | 0.840 |
| 25 to 40 | 0.720 |
| 40 to 55 | 0.627 |
| 55 to 70 | 0.547 |
| 70 to 85 | 0.453 |
| 85 to 100 | 0.373 |

*Percent:*

0: No rock—few small pebbles.
0 to 25: Minor small rocks.
25 to 40: Major rocks and some small boulders.
40 to 55: Major rocks and some small and large boulders.
55 to 70: Minor small and large boulders with frequent occurrence.
70 to 85: Large boulders with frequent occurrence.
85 to 100: Large boulders with frequent to almost continuous occurrence.

Productivity factors for offshore surf-zone bottom conditions to be applied to linear feet of submerged pipelay per day are as follows if pipeline is to be buried:

BOTTOM CONDITION PRODUCTIVITY FACTORS
OFFSHORE SUBMERGED SURF-ZONE PIPELAY

| Description of Bottom | Productivity Factor |
|---|---|
| Gradually Sloping Sandy Loam | 1.000 |
| Gradually Sloping Muck and Mud | 0.850 |
| Soft Rock | 0.750 |
| Medium Soft Rock | 0.650 |

*Gradually Sloping Sandy Loam:* Sand and dirt easily removed by dredge bucket operation, backhoe or jetting.

*Gradually Sloping Muck and Mud:* Muck and mud removable with dredge bucket, backhoe or jetting.

*Soft Rock:* Rock fairly easily removed with dredge buckets or backhoe.

*Medium Soft Rock:* Rock somewhat more difficult to remove with dredge buckets or backhoe.

If blasting is required, see Section Five—DREDGING, BLASTING AND JETTING.

# EXAMPLE—APPLICATION OF FACTORS

The following example shows how to apply the labor, terrain, rock and bottom factor tables in this section to correctly arrive at the total linear feet of pipelay one might expect to accomplish in a work day as outlined.

For our example we assume that we have three thousand (3000) linear feet of 20-inch by ⅜-inch wall pipe to be installed in a trench on land and two thousand (2000) linear feet of concrete weight coated 20-inch by ⅜-inch wall pipe to be installed in a trench in the offshore surf-zone area.

In accordance with our schedule we estimate that the on-land portion can be installed working a ten (10) hour day and that the submerged surf-zone portion, due to the expensive cost of offshore equipment, will be installed utilizing a double shift working twelve (12) hours each for a twenty-four (24) hour work day.

We further assume that, due to the fact that we are going to continue the submerged pipelay for several miles offshore to a platform, we will use the land fabricate—pull barge method of installation for the surf area submerged pipelay.

First we consider the on-land portion of the pipe. From the preceding "On-Land, Near-Shore, Underground Pipelay Table" (page 68), we determine that two thousand nine hundred (2,900) linear feet of 20-inch by ⅜-inch wall pipe can be laid in a trench, in level sandy loam during a ten (10) hour work day.

From the labor analyzation table, and the example of labor productivity (pages xi-xiv), to this manual, it has been determined that our labor productivity is equal to sixty-five (65) percent.

# EXAMPLE—APPLICATION OF FACTORS (continued)

Let us assume that the total on-land pipeline is not to be laid in total level sandy loam and that the following terrain and rock factors must be applied:

EXAMPLE TERRAIN FACTORS

| Description of Terrain | Terrain Factor | Linear Feet of Pipe | Use Factor |
|---|---|---|---|
| Level Sandy Loam | 1.000 | 2000 | 2000.0 |
| Sandy Loam With Sand Dunes | 0.950 | 300 | 285.0 |
| Muck and Mud | 0.800 | 200 | 160.0 |
| Minor Up and Down Grade Changes | 0.750 | 400 | 300.0 |
| Steep Upgrade Changes | 0.500 | 100 | 50.0 |
| Total | — | 3000 | 2795.0 |

$$\text{Terrain Factor} = \frac{2795.}{3000.} = 0.932$$

EXAMPLE ROCK FACTORS

| Percent of Rock | Rock Factor | Linear Feet of Pipe | Use Factor |
|---|---|---|---|
| 0 | 1.000 | 2500 | 2500.0 |
| 0 to 25 | 0.840 | 100 | 84.0 |
| 25 to 40 | 0.720 | 200 | 144.0 |
| 40 to 55 | 0.627 | 100 | 63.0 |
| 55 to 70 | 0.547 | 100 | 55.0 |
| Total | — | 3000 | 2846.0 |

$$\text{Rock Factor} = \frac{2846.}{3000.} = 0.949$$

# EXAMPLE—APPLICATION OF FACTORS (continued)

Based on the preceding assumption, the average rate of on-land pipelay per ten (10) hour day is equal to:

Linear Feet Per Day × Labor Factor × Terrain Factor × Rock Factor

or

3000 L.F. × 0.65 × 0.932 × 0.949 = 1725 L.F. Per Day

On-land piping on supports can be estimated the same way, using the factors that may apply.

Next, for our example, we consider the surf-zone, concrete weight coated, submerged pipelay. From the pipelay table on page 70, we determine that two thousand five hundred ninety (2590) linear feet of 20-inch by ⅜-inch wall pipe can be laid in a trench on a gradually sloping sandy loam bottom during a double work shift of twelve (12) hours each or a twenty-four (24) hour period.

## EXAMPLE—APPLICATION OF FACTORS (continued)

Assume that labor productivity is sixty-five (65) percent, as previously explained and outlined.

Let us assume that the total offshore piping is not to be installed in total gradually sloping sandy loam and that the following bottom condition factors must be applied:

| Description of Bottom | Bottom Factor | Linear Feet of Pipe | Use Factor |
|---|---|---|---|
| Gradually Sloping Sandy Loam .......... | 1.000 | 1200 | 1200.0 |
| Soft Rock .............................. | 0.750 | 200 | 150.0 |
| Medium Soft Rock ...................... | 0.650 | 100 | 65.0 |
| Total .................................. | — | 2000 | 1415.0 |

$$\text{Bottom Factor} = \frac{1415}{2000} = 0.708$$

Average rate of offshore, surf-zone pipelay per twenty-four (24) hour day, based on the preceding assumption is equal to:

$$\text{Linear Feet Per Day} \times \text{Labor Factor} \times \text{Bottom Factor}$$

or

$$200 \text{ L.F.} \times 0.65 \times 0.708 = 920 \text{ L.F. Per Day}$$

The direct installation cost per linear foot of pipelay can be obtained by the following formula:

$$\text{Direct Cost Per Foot} = \frac{\text{Cost Per Day of Spreads}}{\text{Linear Feet Per Day of Lay}}$$

# ONSHORE CLEARING AND GRADING RIGHT-OF-WAY

## General Notes

When estimating a total time frame for the following labor crew and equipment spread, the following items, as they may apply to a given project, should be given consideration:

1) Removal of trees, brush and stumps.
2) Grubbing and removal of stumps that are in the way of the ditch.
3) Disposal of all debris, including method of disposal and length of haul.
4) Clear area spoil a sufficient distance from the ditchline to insure that the spoil bank from the ditching operations will not fall in any foreign material that might become mixed with the excavated spoil.
5) Cutting of merchantable timber into standard length and stacked along the right-of-way for disposition by others (if specifically required by the right-of-way agreement).
6) Provide temporary walks, passageways, fences or other structures so as not to interfere with traffic.
7) Provide sufficient and proper lighting where required.
8) Provide guards where required.
9) Preserve all trees, shrubs, hedges and lawns where required.
10) Grade irregularities where required.
11) Preserve topsoil for replacement, through improved areas, to its original position.
12) Proper grading of the terrain to allow passage of loaded trucks and equipment-hauling vehicles so ditching operations can be performed efficiently.
13) Protect and preserve existing drainage facilities.
14) Protect any existing structures or pipelines.
15) Protect any utility or telephone lines and keep them in service.
16) Install gates and fences where necessary.

# ONSHORE CLEAR AND GRADE RIGHT-OF-WAY

**Labor Crew**

| Personnel Description | NUMBER OF MEN FOR | | | | | | | | | | | |
|---|---|---|---|---|---|---|---|---|---|---|---|---|
| | 50 Linear Feet Width | | | | 80 Linear Feet Width | | | | 100 Linear Feet Width | | | |
| | L | M | MH | H | L | M | MH | H | L | M | MH | H |
| Foreman ......... | 1 | 1 | 1 | 1 | 1 | 1 | 1 | 1 | 1 | 1 | 1 | 1 |
| Operator ......... | 1 | 1 | 1 | 1 | 1 | 1 | 2 | 2 | 2 | 2 | 3 | 3 |
| Mechanic ........ | 1 | 1 | 1 | 1 | 1 | 1 | 1 | 1 | 1 | 1 | 1 | 1 |
| Swamper ......... | 1 | 1 | 1 | 1 | 1 | 1 | 2 | 2 | 2 | 2 | 2 | 2 |
| Truck Driver ..... | 2 | 2 | 2 | 2 | 2 | 2 | 3 | 3 | 3 | 3 | 3 | 3 |
| Laborer .......... | 4 | 4 | 6 | 6 | 8 | 8 | 10 | 10 | 12 | 12 | 14 | 14 |
| Total Crew ......... | 10 | 10 | 12 | 12 | 14 | 14 | 19 | 19 | 21 | 21 | 24 | 24 |

Above total crew should be ample for clearing and grading one-half (½) mile of right-of-way per ten (10) hour day for the widths and conditions outlined.

Crew spread includes cutting, stacking or piling, loading and hauling a round trip distance of two (2) miles.

If burning is necessary or permitted, substitute fire tenders for dump truck drivers. See clearing and grading equipment spread for number of dump trucks.

*Code Description:*

L = Light—light brush and grass, no trees.
M = Medium—considerable brush of larger size.
MH = Medium Heavy—large brush and small trees.
H = Heavy—much small brush, many small trees and occasional large trees.

# ONSHORE CLEARING AND GRADING RIGHT-OF-WAY

**Equipment Spread**

| Equipment Description | NUMBER OF UNITS FOR | | | | | | | | | | | |
|---|---|---|---|---|---|---|---|---|---|---|---|---|
| | 50 Linear Feet Width | | | | 80 Linear Feet Width | | | | 100 Linear Feet Width | | | |
| | L | M | MH | H | L | M | MH | H | L | M | MH | H |
| D8 Tractor With Dozer | 1 | 1 | 1 | 1 | 1 | 1 | 1 | 1 | 1 | 1 | 2 | 2 |
| D7 Tractor With Dozer | 0 | 0 | 0 | 0 | 0 | 0 | 1 | 1 | 1 | 1 | 1 | 1 |
| Truck—2½ Ton Dump | 1 | 1 | 1 | 1 | 1 | 1 | 2 | 2 | 2 | 2 | 2 | 2 |
| Truck—Pick-up ..... | 1 | 1 | 1 | 1 | 1 | 1 | 1 | 1 | 1 | 1 | 1 | 1 |
| Ripper or Brush Rake | 1 | 1 | 1 | 1 | 1 | 1 | 1 | 1 | 1 | 1 | 2 | 2 |

Above equipment spread should be ample for clearing and grading one-half (½) mile of right-of-way per ten (10) hour day for the widths and conditions outlined.

Haul trucks are based on round-trip haul of two (2) miles. If brush and trees are to be burned on site, delete above dump trucks.

Small tools such as saws, axes, etc., must be added as required for the individual job.

*Code Description:*

L = Light—light brush and grass, no trees.
M = Medium—considerable brush of larger size.
MH = Medium Heavy—large brush and small trees.
H = Heavy—much small brush, many small trees and occasional large trees.

# ONSHORE PIPELINE LAY-OUT

## Labor Crew

| Personnel Classification | Number of Men Required |
|---|---|
| Engineer—Surveyor | 1 |
| Rod Man | 1 |
| Chain Man | 1 |
| Laborer | 2 |
| Truck Driver | 1 |
| Total Crew | 6 |

## Equipment and Tool Spread

| Equipment and Tool Description | Number of Units Required |
|---|---|
| Transit | 1 |
| Flag Pole | 1 |
| Tape Chains | 1 |
| Ten Pound Sledge Hammer | 1 |
| Axe | 1 |
| Pick-up Truck | 1 |

Above labor crew, equipment and tools are ample for layout and staking of up to three thousand (3000) linear feet of near-shore pipeline per ten (10) hour day.

Trench bottom grades can be shot with transit; however, it may be desirable to accomplish this with the use of a level. If level is to be used, add this to the above equipment spread.

# ONSHORE UNLOADING, HANDLING, HAULING AND STRINGING PIPE

## General Notes

In estimating the time requirements for the following labor crew and equipment spread, consideration should be given to the following items should they be related to the particular project at hand:

1) Obtain hauling permits if required by federal, state and/or local law.
2) Offload pipe, valves, fittings, pipe materials, casings, etc., at point of delivery if required, and haul to storage facilities.
3) Secure storage sites.
4) Unload and store on racks if required.
5) Check material for damage when received.
6) Load, haul and string pipe and materials along right-of-way.
7) When handling pre-coated pipe, protect coating by use of padded chains or slings.

# ONSHORE UNLOADING, HANDLING, HAULING AND STRINGING PIPE

### Labor Crew

| Personnel Classification | Number of Men Required |
|---|---|
| Foreman | 1 |
| Operator | 2 |
| Oiler | 2 |
| Truck Driver | 2 |
| Hooker | 2 |
| Laborer | 4 |
| Total Crew | 13 |

### Equipment Spread

| Equipment Description | Number of Units Required |
|---|---|
| Crane | 1 |
| D7 Side Boom | 1 |
| Truck With Pipe Trailer | 1 |
| Truck—Pick-up | 1 |

Above labor crew and equipment spread is ample for handling, hauling, unloading and stringing pipe along a right-of-way that is accessible to vehicles involved.

Length of haul and size and quantity of pipe will determine time above labor crew and equipment spread will be required.

# ONSHORE AND OFFSHORE DITCHING AND TRENCHING

## General Notes

In determining a time frame for application of the following labor crews and equipment spreads, the following should be given consideration should they apply to the particular project:

*On-Shore:*

1) Ditching or trenching for buried pipelines should be in accordance with the following table of minimum width and coverage for all soil formations.

| Nominal Pipe Size Inches | Minimum Width Inches | Normal Minimum Coverage Inches |
|---|---|---|
| 4 | 22 | 30 |
| 6 | 26 | 30 |
| 8 | 26 | 30 |
| 10 | 26 | 30 |
| 12 | 30 | 30 |
| 14 | 32 | 30 |
| 16 | 36 | 30 |
| 18 | 38 | 30 |
| 20 | 40 | 30 |
| 24 | 44 | 30 |
| 30 | 50 | 30 |
| 36 | 52 | 36 |
| 42 | 58 | 36 |

2) In rock, cut ditches at least six (6) inches wider.
3) If dirt-filled benches are used, ditch should be excavated deeper to obtain proper coverage.

# ONSHORE AND OFFSHORE DITCHING AND TRENCHING

## General Notes (continued)

4) Trench should be excavated to greater depth for proper installation of the pipe where the topography is such to warrant this.
5) Repair any damage to and maintain existing natural or other drainage facilities.
6) Do not open ditch too far in advance of pipelay crew.
7) Obtain permits for blasting.
8) When blasting, use extreme caution and protection.
9) Clean-up blasted rock to prevent damage to coated pipe.

*Offshore:*

1) Cut ditch sufficiently wide and deep to compensate for wash fills due to tides and water movements.
2) Do not open trench until pipe is ready to be placed.
3) When backfill is required, proceed with this operation as soon as pipe is in place.

# ONSHORE DITCHING AND TRENCHING

**Labor Crew**

| Personnel Description | Number of Men Required |
|---|---|
| Foreman | 1 |
| Operator | 3 |
| Oiler | 3 |
| Swamper | 1 |
| Grader | 1 |
| Operator—Jackhammer | 2 |
| Operator—Drills | 2 |
| Powder Man | 1 |
| Truck Driver | 4 |
| Laborer | 6 |
| Total Crew | 24 |

Manpower will vary, depending on length, width and type of excavation, type of terrain, and soil conditions.

The above labor crew is based on conditions outlined on pages 77-80 and should be the maximum required to open ditch far enough in advance of the pipelaying crews so as not to delay pipelay operations.

If demolition of rocks and boulders is unnecessary, or if no blasting is required, delete the craftsmen for these operations from the above crew.

Water height permitting, the above crew can be utilized for excavation in the surf-zone area at low tide.

# ONSHORE DITCHING AND TRENCHING

## Equipment Spread

| Equipment Description | Number of Units Required |
|---|---|
| Backhoe With ⅝ Cy. to 1 Cy. Bucket | 1 |
| Tractor With Dozer Blade | 1 |
| Tractor With Tow—D6 or D7 | 1 |
| Air Compressor—365 to 500 CFM. | 1 |
| Jackhammer—Light to Heavy | 2 |
| Drills—Wagon or Train—4" to 5½" | 4 |
| Water Pumps—2" to 6" | 2 |
| Trucks—Dump—5 to 12 Cy. | 3 |
| Trucks—Pick-up—½ to ¾ Ton | 2 |

Size of above equipment may vary, depending on length, width and type of excavation, type of terrain, and soil conditions.

The above equipment spread is based on the conditions outlined on pages 77-80 and should be the maximum required to open ditch far enough in advance of the pipelaying crews so as not to delay pipelay operations.

If demolition of rocks and boulders is unnecessary, or no blasting is required, delete the equipment for these operations from the above spread.

Water height permitting, the above spread can be utilized for excavation in the surf-zone area at low tide.

# OFFSHORE DITCHING AND TRENCHING

## Labor Crew

| Personnel Description | Number of Men Required |
|---|---|
| Barge/Ditch Foreman | 1 |
| Mechanic | 1 |
| Anchor Operator | 1 |
| Backhoe Operator | 1 |
| Skiff Operator | 1 |
| Oiler | 1 |
| Laborer/Swamper | 4 |
| Truck Driver | 1 |
| Total Crew | 11 |

## Equipment Spread

| Equipment Description | Number of Units Required |
|---|---|
| Tug—165 HP. (Fully Manned) | 1 |
| Material Barge | 1 |
| Skiff With Motor | 1 |
| Crane/Backhoe | 1 |
| Carry All | 1 |
| Walkie Talkie | 2 |

Above labor and equipment spreads are ample for ditching operations in the offshore surf-zone area in sand, mud and light clay type soils.

Length of time required for the above crew and spread will be determined by the type of soil, width and depth of ditch, and length of run.

Where feasible the use of a land spread should be used in the surf-zone area at low tide.

If blasting of coral or rock is required, see Section Five—DREDGING, BLASTING AND FITTING.

A fully manned dredge can be substituted for the above if desired.

If excavated material is to be disposed of at sea, additional equipment, fully manned for this operation, must be added.

Labor crew and equipment spread similar to the above can be used for backfilling.

# ONSHORE BENDING OPERATIONS

## General Notes

In estimating the time requirements for bending operations, consideration should be given to the following items:

1) Bending machine should be of proper size and design.
2) All bends should be cold bends without wrinkles.
3) Curvature of bend should be distributed uniformly throughout as great a length of the pipe as possible.
4) Bends should be used for difference in elevation of the surface of the ground as well as for alignment.
5) Bends should not exceed one and one-half degrees (1½°) per diameter length of pipe.
6) Bend should be made no nearer than six (6) feet from the end of the joint of pipe. When pipe is double-jointed before bending, bends should not be closer to the weld than one (1) pipe diameter. If, however, it should be necessary to bend pipe through a circumferential weld, the weld should be x-rayed after bending.

# ONSHORE BENDING OPERATIONS

### Labor Crew

| Personnel Classification | Number of Men Each |
|---|---|
| Foreman | 1 |
| Engineer | 1 |
| Engineer Helper | 1 |
| Operator—Tractor | 1 |
| Operator—Bending Machine | 1 |
| Operator—Bending Mandrel | 1 |
| Oiler | 1 |
| Pipefitter | 2 |
| Laborer | 4 |
| X-Ray Technician | 1 |
| Total Crew | 14 |

### Equipment Spread

| Equipment Description | Number of Units Each |
|---|---|
| Pipe Bending Machine | 1 |
| Pipe Bending Mandrel | 1 |
| Dent Remover | 1 |
| Air Compressor 365 CFM | 1 |
| Tractor W/Side Boom D6 or D7 | 1 |
| Transit | 1 |
| Truck—Pick-up | 1 |
| X-Ray Equipment | 1 |

Above labor crew and equipment spread are ample for minor bending operations of all sizes.

Size of pipe will dictate size of bending equipment to be used.

When possible, bends should be shipped to the job prefabricated.

# ONSHORE ALIGNING AND WELDING OPERATIONS

## General Notes

In keeping with the average pipelay table (page 46), consideration should be given to the following items for aligning and welding operations:

1) The labor crew and equipment spread is based on welds made by the manual shielded metal-arc method.
2) Welding machines should be of the size and type suitable for the work.
3) Welding machines are to be operated within the amperage and voltage ranges recommended for each size and type of electrode.
4) Remove all foreign matter from the beveled ends before welding. Clean with hand or power tools.
5) Pipe ends to have proper bevels for the work intended.
6) The alignment of abutting pipe ends should be such as to minimize the offset between pipe surface. Hammering of pipe to obtain proper line-up should be held to a minimum.
7) Internal line-up clamps should be used where possible. Use external line-up clamps when it is impractical or impossible to use an internal line-up clamp.
8) If roll welding is used, add skids or structural framework and required roller dollies to the equipment spread.
9) Use a minimum of root bead, filler bead and finish bead.
10) Brush and clean the finished weld.
11) Only qualified welders should be used.
12) Where required, apply heat treatment—preheating and postheating.
13) Conduct radiographic inspection of welds where required.

# ONSHORE ALIGNING AND WELDING OPERATIONS

**Labor Crew**

| Personnel Description | MEN EACH FOR PIPE SIZES | | |
|---|---|---|---|
| | 4" to 14" | 16" to 24" | 20" to 24" |
| Foreman | 1 | 2 | 2 |
| Welder—Stringer | 2 | 2 | 2 |
| Welder—Line | 2 | 4 | 6 |
| Welder—Helper | 4 | 6 | 8 |
| Clamp Man | 2 | 2 | 2 |
| Stabber | 1 | 1 | 1 |
| Spacer | 1 | 1 | 1 |
| Buffer | 2 | 2 | 2 |
| Operator | 3 | 3 | 3 |
| Oiler | 3 | 3 | 3 |
| Swamper | 1 | 1 | 1 |
| Truck Driver | 2 | 2 | 2 |
| Laborer | 4 | 5 | 6 |
| X-Ray Technician | 1 | 1 | 1 |
| Total Crew | 29 | 35 | 40 |

The above labor crew is the maximum that should be required for aligning and welding of pipe as outlined in the average pipelay table and the example application of factors on pages 72-80.

Above manpower is based on aligning and welding up to three thousand (3000) linear feet of pipeline. Above spread should be adjusted upward for larger projects.

It is not the intent of the above labor crew to provide manpower requirements for piping of tank farms, blending facilities or process facilities.

# ONSHORE ALIGNING AND WELDING OPERATIONS

**Equipment Spread**

| Equipment Description | NUMBER OF UNITS PER SIZES | | |
|---|---|---|---|
| | 4" to 14" | 16" to 24" | 30" to 42" |
| 300 Amp. Welding Machine ....... | 4 | 6 | 8 |
| Beveling and Cutting Rig ......... | 1 | 1 | 1 |
| D7 Side Boom .................... | 2 | 2 | 2 |
| D6 Tow Tractor ................... | 1 | 1 | 1 |
| Buffing Outfit .................... | 2 | 2 | 2 |
| Line-Up Clamps .................. | 2 | 2 | 2 |
| Truck—2 Ton Winch ............. | 1 | 1 | 1 |
| Truck—5 Ton .................... | 1 | 1 | 1 |
| Truck—Pick-up .................. | 1 | 1 | 1 |
| Carry All or Bus .................. | 1 | 1 | 1 |
| X-Ray Equipment—Set Up ........ | 1 | 1 | 1 |

The above equipment spread is the maximum that should be required for aligning and welding of pipe as outlined in the average pipelay table and the example application of factors on pages 72-80.

Above equipment spread is based on aligning and welding up to three thousand (3000) linear feet of pipeline. Above spread should be adjusted upward for larger projects.

It is not the intent of the above spread to provide equipment requirements for piping of tank farms, blending facilities or process facilities.

# ONSHORE, LAND SHOVE, SUBMERGED SURF-ZONE PIPE INSTALLATION ALIGNING, WELDING AND SHOVING OPERATIONS

**Labor Crew**

| Personnel Description | NUMBER OF MEN FOR SIZES | | |
|---|---|---|---|
| | 4″ to 14″ | 16″ to 24″ | 30″ to 42″ |
| Foreman | 1 | 1 | 1 |
| Welder | 4 | 6 | 8 |
| Welder Helper | 4 | 6 | 8 |
| Spacer | 1 | 1 | 1 |
| Stabber | 1 | 1 | 1 |
| Buffer | 2 | 4 | 6 |
| Operator | 5 | 6 | 6 |
| Oiler | 3 | 4 | 4 |
| Swamper | 1 | 2 | 2 |
| Pot Fireman | 1 | 1 | 1 |
| Laborer | 6 | 8 | 10 |
| Truck Driver | 3 | 3 | 3 |
| X-Ray Technician | 1 | 1 | 1 |
| Total Crew | 33 | 44 | 52 |

This labor crew is for installing piping from the shoreline outward to sea and can be used for this type of installation without the back-up of a lay barge to pull the line, assuming distance and surf-zone conditions will permit this type of operation.

The above labor crew should be added to the pipelay barge labor crew if the land-fabricate barge pull method is used.

See section 4, "Ocean and Sea Pipelines" page 153, for pipelay barge labor crew.

# ONSHORE, LAND SHOVE, SUBMERGED SURF-ZONE PIPE INSTALLATION ALIGNING, WELDING AND SHOVING OPERATIONS

**Equipment Spread**

| Equipment Description | NUMBER OF UNITS FOR SIZES | | |
|---|---|---|---|
| | 4″ to 14″ | 16″ to 24″ | 30″ to 42″ |
| Crane—50 Ton± | 1 | 1 | 1 |
| Side Boom—D6 or D7 | 1 | 2 | 2 |
| Clam or Drag Bucket—¾ Cy | 1 | 1 | 1 |
| Welding Machine—300 Amp | 4 | 6 | 8 |
| Generator—75 KW | 2 | 3 | 3 |
| Air Compressor—165 CFM | 1 | 1 | 1 |
| Hoist—Pull | 1 | 1 | 1 |
| Hoist—Hold Back | 1 | 1 | 1 |
| Line Up Clamps | 2 | 2 | 2 |
| Concrete Mixer | 1 | 1 | 1 |
| Tar Kettle | 1 | 1 | 1 |
| Jet Pump 6″ | 1 | 1 | 1 |
| Launch Shoes | 1 | 1 | 1 |
| Skiff With Motor | 1 | 1 | 1 |
| Winch Truck—2 Ton | 1 | 1 | 1 |
| Truck—Pick-up | 1 | 1 | 1 |
| Carry All or Bus | 1 | 1 | 1 |
| Holiday Detector | 1 | 1 | 1 |
| X-Ray Set-up | 1 | 1 | 1 |
| Radio—Walkie Talkie | 2 | 2 | 2 |

This equipment spread is for installing piping from the shoreline outward to sea and can be used for this type of installation without the back-up of a lay barge to pull the line, assuming distance and surf-zone conditions will permit this type of operation.

This equipment spread should be added to the pipelay barge equipment spread if the land-fabricate barge pull method is used.

Onshore concrete deadmen or anchor system for anchoring hoist is not included above and must be added to fit the needs of the individual project.

See Section 4, "Ocean and Sea Pipelines" page 154 for pipelay barge equipment spread.

# ONSHORE, LAND PULL, SUBMERGED SURF-ZONE PIPE INSTALLATION ALIGNING AND PULLING OPERATIONS

**Labor Crew**

| Personnel Description | NUMBER OF MEN FOR SIZES | | |
|---|---|---|---|
| | 4″ to 14″ | 16″ to 24″ | 30″ to 42″ |
| Foreman | 1 | 1 | 1 |
| Welder | 2 | 4 | 4 |
| Welder Helper | 1 | 2 | 2 |
| Operator | 4 | 5 | 5 |
| Oiler | 4 | 5 | 5 |
| Swamper | 1 | 2 | 2 |
| Laborer | 6 | 8 | 10 |
| Truck Driver | 3 | 3 | 3 |
| Total Crew | 22 | 30 | 32 |

This labor crew is for onshore assisting by pulling. Lay barge-fabricated pipe pushed from a position offshore beyond the surf-zone.

The above labor crew should be added to the pipelay barge labor crew if the lay barge-fabricate land pull method is used.

See Section 4, "Ocean and Sea Pipelines" page 153 for pipelay barge labor crew.

# ONSHORE, LAND PULL, SUBMERGED SURF-ZONE PIPE INSTALLATION ALIGNING AND PULLING OPERATIONS

**Equipment Spread**

| Equipment Description | NUMBER OF UNITS FOR SIZES | | |
|---|---|---|---|
| | 4″ to 14″ | 16″ to 24″ | 30″ to 42″ |
| Crane—50 Ton± | 1 | 1 | 1 |
| Side Boom—D6 or D7 | 1 | 2 | 2 |
| Clam or Drag Bucket ¾ Cy | 1 | 1 | 1 |
| Welding Machine | 1 | 2 | 2 |
| Generator—75 KW | 1 | 2 | 1 |
| Air Compressor—165 CFM | 1 | 1 | 1 |
| Hoist—Pull | 1 | 1 | 1 |
| Jet Pump—6″ | 1 | 1 | 1 |
| Skiff With Motor | 1 | 1 | 1 |
| Winch Truck—2 Ton | 1 | 1 | 1 |
| Truck—Pick-up | 1 | 1 | 1 |
| Carry All or Bus | 1 | 1 | 1 |
| Radio—Walkie-Talkie | 2 | 2 | 2 |

This equipment spread is for onshore assisting by pulling lay barge-fabricated pipe pushed from a position offshore beyond the surf-zone.

The above equipment spread should be added to the pipelay barge equipment spread if the method of lay barge-fabricate land pull method is used.

Onshore concrete deadmen or anchor system for anchoring hoist are not included above and must be added to fit the needs of the individual project.

See Section 4, "Ocean and Sea Pipelines" page 154 for pipelay barge equipment spread.

# ONSHORE CLEANING, PRIMING, COATING AND WRAPPING OPERATIONS

## General Notes

Consideration should be given to the items listed below when estimating the time requirements for the labor crews and equipment spreads on pages 102-105:

1) Clean and remove all dirt, mill scale, rust, welding scale and all other foreign material.
2) Some or all of the following operations may be required in applying the coating and wrapping the pipe or joints:
   i) Priming
   ii) Enamel finish
   iii) Fiber glass wrap
   iv) Enamel coat No. 2
   v) Asbestos felt wrap
   vi) Fiber glass wrap No. 2
   vii) Enamel coat No. 3
   viii) Kraft wrapping
   ix) Unbonded wrap
   x) Whitewash
   xi) Wood lagging
3) Always handle coated pipe with equipment designed to prevent damage to the coating.
4) Where pipe is to be laid in trench excavated in rock, pad trench to prevent damage to pipe coating.
5) Check all coatings for holidays with a high-voltage electric holiday.

# FIELD ONSHORE CLEANING, PRIMING, COATING AND WRAPPING OPERATIONS FOR TOTAL PIPELINE

**Labor Crew**

| Personnel Description | Number of Men Required |
|---|---|
| Foreman | 1 |
| Operator—Equipment | 3 |
| Operator—Cleaning Machine | 1 |
| Operator—Coating Machine | 1 |
| Operator—Dope Pot | 1 |
| Pot Fireman | 1 |
| Oiler | 2 |
| Paper Lather | 2 |
| Swamper | 2 |
| Truck Driver | 1 |
| Laborer | 6 |
| Total Crew | 21 |

The above labor crew is capable of cleaning, priming, coating and wrapping much more pipe than is shown in the average pipelay table (page 46). However, this crew is necessary for operation of the equipment required if the line is to be coated and wrapped in the field.

If pipe is to be received pre-coated and wrapped, see table on page 103 for field cleaning, priming, coating and wrapping of joints only.

**Equipment Spread**

| Equipment Description | Number of Units Required |
|---|---|
| Cleaning Machine | 1 |
| Coating and Wrapping Machine | 1 |
| D7—Side Boom | 2 |
| D6—Tow Tractor | 1 |
| Holiday Detector | 1 |
| Truck—3 Ton | 1 |
| Truck—Pick-up | 1 |
| 6 KW Generator | 2 |

This equipment spread allows for cleaning, priming, coating and wrapping much more pipe than is shown in the average pipelay table (page 46). However, this equipment is necessary if the pipe is coated and wrapped in the field.

If pipe is to be received pre-coated and wrapped, see table on page 103 for equipment required for field cleaning, priming, coating and wrapping of joints only.

# TOTAL FIELD ONSHORE CLEANING, PRIMING, COATING AND WRAPPING OPERATIONS OF JOINTS ONLY

**Labor Crew**

| Personnel Description | Number of Men Required |
|---|---|
| Foreman | 1 |
| Operator | 1 |
| Oiler | 1 |
| Buffer | 2 |
| Dye, Enamel and Wrappers | 4 |
| Pot Fireman | 1 |
| Laborer | 2 |
| Truck Driver | 1 |
| Total Crew | 13 |

**Equipment Spread**

| Equipment Description | Number of Units Required |
|---|---|
| D7—Side Boom | 1 |
| 3 Bbl Tar Kettle | 1 |
| Holiday Detector | 1 |
| 6 KW Generator | 2 |
| Buffer | 3 |
| Pick-up Truck | 1 |

The above labor crew and equipment spread will cover cleaning, priming, coating and wrapping of pipe joints only, for the quantities of pipelay listed in the average pipelay table on page 68-69.

If total pipeline is to be field coated and wrapped, see tables on page 102.

# SANDBLAST AND PAINT PIPE

## Labor Crew

| Personnel Description | Number of Men |
|---|---|
| Foreman | 1 |
| Blasters and Painters | 3 |
| Pot Man or Helper | 4 |
| Operator | 1 |
| Truck Driver | 1 |
| Total Crew | 10 |

## Equipment Spread

| Equipment Description | Number of Units |
|---|---|
| Sand Pot and Blast Equipment | 1 |
| Print Spray Equipment | 1 |
| Air Compressor—600 CFM | 1 |
| Truck—Pick-up | 1 |

The above labor crew and equipment spread should be ample for cleaning (by blasting) and painting pipe installed on supports or racks and should be capable of blasting twelve hundred (1200) square feet and painting one thousand (1000) square feet of pipeline per ten (10) hour day.

# ONSHORE LOWERING PIPE IN TRENCH

## General Notes

The following operations should be considered when lowering pipe into a trench:

1) Remove rocks and other objects from trench that could damage pipe.
2) Allow slack in line to properly fit ditch.
3) Use only canvas slings to lower pipe into trench.
4) Use holiday detector behind the last lowering in belt at all times.
5) Where it is necessary to pull or drag sections of pipe, protect coating.
6) Keep open ends of pipe closed at all times with "night-caps."
7) Should dents, buckles or defects be found in pipe, replace same.

# ONSHORE LOWERING OF PIPE IN TRENCH

**Labor Crew**

| Personnel Description | MEN EACH FOR PIPE SIZES | |
|---|---|---|
| | 4" to 20" | 24" to 42" |
| Foreman | 1 | 1 |
| Operator | 2 | 3 |
| Oiler | 2 | 3 |
| Swamper | 1 | 1 |
| Laborers | 4 | 6 |
| Truck Driver | 1 | 1 |
| Total Crew | 11 | 15 |

**Equipment Spread**

| Equipment Description | NUMBER OF UNITS FOR PIPE SIZES | |
|---|---|---|
| | 4" to 20" | 24" to 42" |
| Side Boom—D7 or D8 | 2 | 3 |
| Drag Line—½ to ¾ Cy | 1 | 1 |
| Water Pump—4" to 6" | 2 | 2 |
| Truck—Pick-up | 1 | 1 |

Above labor crew and equipment spread should be ample for lowering pipe, of the sizes listed, into a pre-excavated trench, based on the pipe quantities listed in the average pipelay table, page 68.

# VALVE INSTALLATION

## Labor Crew

| Personnel Description | Number of Men |
|---|---|
| Foreman | 1 |
| Pipefitter/Welder | 6 |
| Pipefitter Helper | 6 |
| Operator | 1 |
| Oiler | 1 |
| Laborer | 4 |
| Truck Driver | 1 |
| Total Crew | 20 |

## Equipment Spread

| Equipment Description | Number of Units |
|---|---|
| Light Crane | 1 |
| Welding Machine—300 Amp | 1 |
| Generator | 1 |
| Flatbed Truck—2 Ton | 1 |
| Pick-up Truck | 1 |

The above labor crew and equipment spread should be ample for the installation of valves in pipelines.

The number of valves to be installed and the valve size, type and location will govern the actual size of crew and equipment spread required.

See page 108 for average hourly time requirements for valve installation.

# VALVE INSTALLATION

**Average Hourly Time Requirements**

| Nominal Pipe Size | FLANGED AND BOLTED VALVES | | | | |
|---|---|---|---|---|---|
| | SERVICE PRESSURE RATING | | | | |
| | 150 Lb. | 300-400 Lb. | 600-900 Lb. | 1500 Lb. | 2500 Lb. |
| 8 | 17. | 19. | 25. | 35. | 40. |
| 10 | 19. | 24. | 34. | 44. | 50. |
| 12 | 27. | 29. | 42. | 55. | 63. |
| 14 | 31. | 35. | 49. | 63. | 73. |
| 16 | 36. | 40. | 56. | 72. | 85. |
| 18 | 41. | 46. | 64. | 85. | 98. |
| 20 | 48. | 54. | 73. | 108. | 110. |
| 22 | 53. | 60. | 83. | 111. | 114. |
| 24 | 58. | 65. | 92. | 121. | 124. |
| 26 | 65. | 71. | 100. | 123. | — |
| 30 | 79. | 86. | 119. | 144. | — |
| 36 | 100. | 108. | 148. | 177. | — |
| 42 | 121. | 130. | 176. | 211. | — |
| | WELD END VALVES | | | | |
| 8 | 8. | 9. | 10. | 11. | 13. |
| 10 | 10. | 10. | 13. | 15. | 17. |
| 12 | 13. | 13. | 17. | 19. | 23. |
| 14 | 15. | 16. | 21. | 24. | 31. |
| 16 | 19. | 20. | 26. | 30. | 38. |
| 18 | 24. | 25. | 33. | 38. | 49. |
| 20 | 27. | 28. | 40. | 46. | 57. |
| 22 | 31. | 32. | 47. | 53. | 65. |
| 24 | 35. | 37. | 54. | 61. | 72. |
| 26 | 40. | 42. | 69. | 75. | 93. |
| 30 | 49. | 51. | 82. | 89. | 109. |
| 36 | 69. | 72. | 113. | 125. | 141. |
| 42 | 101. | 105. | 165. | 180. | 194. |

Above manhours are for total valve installation including loadout, handle, haul, flange welds, bolt-ups and valve end weld where required.

See preceding table for suggested valve installation labor crew and equipment spread.

# ONSHORE CLEANING AND TESTING OF PIPELINES

## General Notes

'he following items should be considered when estimating the time requirements for the ıbor crew and equipment spread on pages 110-111:

1) All pressure testing should be conducted with due regard for the public.
2) Installation of fill line and filling.
3) Sectionalized testing may be required depending on limitations of the water supply or to limit the degree of public exposure.
4) Pretesting major road crossings, railroad crossings and prefabricated sections prior to installation may be necessary or desirable.
5) Dewatering of lines with the use of scrapers or spheres.
6) Should defects be found, repair or replace same.
7) Provide radio communication equipment to control and coordinate all phases of testing.

# CLEANING AND TESTING PIPELINES

**Labor Crew**

| Personnel Description | Number of Men |
|---|---|
| Foreman | 1 |
| Operator | 1 |
| Oiler | 1 |
| Welder | 6 |
| Welder Helper | 6 |
| Pipefitter | 6 |
| Pipefitter Helper | 6 |
| Swamper | 1 |
| Laborer | 6 |
| Truck Driver | 2 |
| Total Crew | 36 |

The above labor crew is ample for pig cleaning and hydrostatic testing of pipelines.

Length of time required to pig and test line will be determined by length of time required to set up pig launcher and receiver and length and size of pipeline to be tested. This must be estimated for the individual project.

# CLEANING AND TESTING PIPELINES

## Equipment Spread

| Equipment Description | Unit | Number of Units |
|---|---|---|
| Light Crane | Each | 1 |
| Pig Launcher | Each | 1 |
| Pig Trap | Each | 1 |
| Pigs | Each | 4 |
| Misc. Pipe, Valves and Fittings | Lot | 1 |
| Anchor Forgings and Tie Downs | Lot | 1 |
| Generator—75 KW | Each | 2 |
| Welding Machine—300 Amp | Each | 6 |
| Air Compressor—165 CFM | Each | 1 |
| Water Pump—6" | Each | 2 |
| Test Pump | Each | 2 |
| Miscellaneous Gauges | Lot | 1 |
| Holiday Detector | Each | 2 |
| Truck—Flatbed | Each | 1 |
| Truck—Pick-up | Each | 1 |
| Carry All or Bus | Each | 1 |
| Radios (Walkie-Talkie) | Each | 2 |

The above equipment spread is ample for pig cleaning and hydrostatic testing of pipelines.

Time requirements, for the above equipment, to pig and test line will be governed by length of time required to set up pig launcher and receiver and length and size of pipeline to be tested. This must be estimated for the individual project.

# ONSHORE BACKFILLING

## General Notes

In estimating the time requirements for the labor crew and equipment spread on page 113, the following items should be considered.

1) Keep backfill as close as possible to lowering-in operation.
2) Hand backfill to six (6) inches above top of pipe.
3) Rock, if available, usually can be used to backfill from six (6) inches above pipe to top of trench.
4) Do not use foreign substances or refuse for backfill material.
5) Place good earth in cultivated areas or where required.
6) Protect backfill from washing.
7) Place riprap when required and where necessary.
8) Cover all bends on the same day they are lowered into trench.

# ONSHORE TRENCH BACKFILLING

**Labor Crew**

| Personnel Description | MEN EACH FOR PIPE SIZES | |
|---|---|---|
| | 4″ to 20″ | 24″ to 42″ |
| Foreman | 1 | 1 |
| Operator | 2 | 3 |
| Oiler | 1 | 2 |
| Swamper | 1 | 2 |
| Laborer | 2 | 4 |
| Total Crew | 7 | 12 |

**Equipment Spread**

| Equipment Description | NUMBER OF UNITS FOR PIPE SIZES | |
|---|---|---|
| | 4″ to 20″ | 24″ to 42″ |
| Tractor—Front End Loader | 1 | 2 |
| Dozer—D6 or D7 | 1 | 1 |
| Truck—Pick-up | 1 | 1 |

Above labor crew and equipment spread should be ample for backfilling of trench after pipe is installed, assuming pipelay at the rate stipulated in the average pipelay table, (page 68).

If joints are exposed for test, additional spread time will be required to return and backfill joints.

If additional fill material is required, add hauling labor and equipment to above spread.

# ONSHORE CLEAN-UP OPERATION

## General Notes

In estimating time requirements for clean-up operations, the following items should be considered:

1) General clean-up of the entire right-of-way.
2) Follow up lowering-in and backfill crews as closely as possible.
3) Remove any rubbish, brush, trees, stumps, paint drums and loose rocks.
4) Restore beach areas to their original contour.
5) Where erosion is possible, install rip-rap or sand bags.
6) Repair damage to or replace fences, terraces, roadways, bridges, culverts, etc.
7) Install pipeline right-of-way markers.

# ONSHORE CLEAN-UP OPERATION

### Labor Crew

| Personnel Description | Number of Men |
|---|---|
| Foreman | 1 |
| Operator | 2 |
| Oiler | 1 |
| Swamper | 2 |
| Truck Driver | 2 |
| Labor | 6 |
| Total Crew | 14 |

### Equipment Spread

| Equipment Description | Number of Units |
|---|---|
| Dozer—D7 or D8 | 2 |
| Front End Loader | 1 |
| Truck—2 Ton | 1 |
| Truck—Dump | 1 |
| Truck—Pick-up | 1 |

Above labor crew and equipment spread should be ample for clean-up, after construction is completed, of debris and miscellaneous materials all in accordance with the average pipelay table (page 68).

The above labor and equipment spread do not include hauling or burning of brush and trees.

# ONSHORE UTILITY OPERATIONS

## General Notes

Utility operations can vary greatly dependent on the obstructions that may lie in the way of the pipeline. Therefore, adjustment of the labor crew and equipment spread on page 117 may be required depending on the operations involved. Some items that may necessitate adjustment of the tables are:

1) Boring under highways, roads, railroads, etc.
2) Placing conductor piping or casings.
3) Break out and remove pavements and repair and replace damage to same.
4) Install and maintain safety precautions as required such as lanterns, barricades, warning signs, etc.

# ONSHORE UTILITY OPERATIONS

### Labor Crew

| Personnel Description | Number of Men |
|---|---|
| Foreman | 1 |
| Mechanic | 1 |
| Operator | 3 |
| Oiler | 2 |
| Air Tool Operator | 2 |
| Swamper | 3 |
| Welder | 2 |
| Welder Helper | 2 |
| Truck Driver | 2 |
| Laborer | 4 |
| Total Crew | 22 |

### Equipment Spread

| Equipment Description | Number of Units |
|---|---|
| Drag Line | 1 |
| Side Side Boom | 1 |
| Dozer | 1 |
| Welding Machine | 1 |
| Boring Machine | 1 |
| Cut and Bevel Outfit | 1 |
| Asphalt Kettle | 1 |
| Air Compressor | 1 |
| Truck—6 × 6 | 1 |
| Truck—2 Ton | 1 |
| Truck—Pick-up | 1 |

Above labor crew and equipment spread are average for installation of casements and miscellaneous borings and other utility type operations that may be involved in the installation of the onshore portion of a pipeline.

The number and type of utility operations will govern the size of crew and spread required for the individual project. The above labor crew and equipment spread must be adjusted to the particular project.

# Section Four

# OCEAN AND SEA PIPELINES

The purpose of this section is to assist in the estimation of time frames, labor crews and equipment spreads required for the installation of offshore submerged concrete weight coated pipelines.

A wide range of variables affect the installation of submerged pipelines; if the proper crews and equipment spreads are utilized, these variables can be reduced to those of labor productivity and wind, wave and current conditions.

Based on one hundred percent productivity, we have included average pipelay and riser installation tables. Also included are percentage efficiency factors to be applied against the pipelay and riser tables, giving due consideration to labor, wind, wave and current conditions. Labor crews and equipment spreads are given for the various operations in the installations of submerged pipelines and for application against the estimated time frames. The labor listed in the following crew tables is for a double shift working a twenty-four (24) hour day, using the equipment spreads full time.

It is assumed that any available deck space on the lay barge will be used for hauling and storing as much of the pipe as possible and that material or cargo barges in the proper size and capacity will be added to the spread as needed.

To find the total direct daily cost of any labor and equipment spread:

1) Determine the daily rate, including all fringes, of each craft or position and multiply each rate by the number of men required for that position.
2) Determine the daily rate of each rigged-up piece of equipment, including all its components.
3) Determine the daily cost of fuel, oil and grease required to run the spread and add this cost to the spread.
4) Determine the cost of small tools and consumable supplies required to support the spread for the project, prorate on a daily basis and add the prorated cost to the daily spread cost.
5) If quartering and catering of personnel are required, estimate the cost per man day, multiply by the number of men in the crew, and add daily rate for the total crew to the spread.
6) Estimate the total costs of (1) through (5) above to arrive at the total direct cost of the spread per working day.

To determine the daily direct cost of a spread for a non-working or stand-by day, simply reduce the working day spread cost by the cost of the fuel, oil and grease and the consumable supplies that would normally be used when working.

# AVERAGE PIPELAY TABLE (Water Depth to 50 Feet)

### Lay or Derrick Barge Fabricate and Install

NUMBER OF SINGLE RANDOM JOINTS OF CONCRETE WEIGHT COATED PIPE PER TWENTY-FOUR (24) HOUR WORK DAY

| | PIPE WALL THICKNESS IN INCHES | | | | | |
|---|---|---|---|---|---|---|
| O.D. Pipe Size Inches | 0.312 through 0.337 | 0.365 through 0.375 | 0.432 through 0.438 | 0.500 through 0.531 | 0.5625 through 0.625 | 0.675 — — |
| 4.500 | 583 | — | 567 | 556 | — | 540 |
| 5.500 | — | — | — | — | — | 420 |
| 5.563 | — | 537 | 530 | 527 | 515 | — |
| 6.625 | — | 507 | 503 | 497 | 485 | — |
| 8.625 | — | 449 | 445 | 440 | 432 | — |
| 10.750 | — | 442 | 438 | 433 | 425 | — |
| 12.750 | — | 397 | 393 | 389 | 381 | — |
| 14.000 | — | 389 | 387 | 385 | 378 | — |
| 16.000 | — | 370 | 367 | 363 | 356 | — |
| 18.000 | — | 344 | 341 | 338 | 331 | — |
| 20.000 | — | 326 | 323 | 320 | 313 | — |
| 22.000 | — | 310 | 307 | 304 | 298 | — |
| 24.000 | — | 297 | 294 | 291 | 285 | — |

The related construction labor and equipment spreads that follow in this section should be ample for installing the above quantities of submerged, concrete weight coated pipe working a twenty-four (24) hour day.

The above single random joints of submerged pipelay are based on assembling concrete weight coated pipe on a lay barge or combination lay/derrick barge properly rigged with equipment necessary for performing the operations involved.

For other sizes, wall thicknesses and water depths see pages 121 through 140.

For unlisted sizes and wall thicknesses extrapolate from above or other average pipelay tables.

If double random lengths of pipe are to be installed, use 65 percent of the above number of joints.

# AVERAGE PIPELAY TABLE (Water Depth to 50 Feet)

**Lay or Derrick Barge Fabricate and Install**

NUMBER OF SINGLE RANDOM JOINTS OF CONCRETE WEIGHT COATED PIPE PER TWENTY-FOUR (24) HOUR WORKING DAY

| O.D. Pipe Size Inches | PIPE WALL THICKNESS IN INCHES | | | | |
|---|---|---|---|---|---|
| | 0.375 | 0.4325 | 0.500 | 0.5625 | 0.625 |
| 26.000 | 288 | 285 | 282 | 279 | 277 |
| 28.000 | 279 | 276 | 273 | 271 | 268 |
| 30.000 | 270 | 267 | 265 | 262 | 259 |
| 32.000 | 260 | 257 | 255 | 252 | 250 |
| 34.000 | 250 | 247 | 245 | 242 | 240 |
| 36.000 | 240 | 237 | 235 | 233 | 231 |
| 40.000 | — | 221 | 218 | 216 | 214 |
| 42.000 | — | 204 | 201 | 200 | 197 |
| 44.000 | — | 153 | 151 | 150 | 148 |
| 48.000 | — | 115 | 113 | 112 | 111 |
| 52.000 | — | 86 | 85 | 84 | 83 |
| 56.000 | — | 64 | 63 | 63 | 62 |
| 60.000 | — | 48 | 47 | 47 | 46 |

The related construction labor and equipment spreads that follow in this section should be ample for installing the above quantities of submerged, concrete weight coated pipe working a twenty-four (24) hour day.

The above single random joints of submerged pipelay are based on assembling concrete weight coated pipe on a lay barge or combination lay/derrick barge properly rigged with equipment necessary for performing the operations involved.

For other sizes, wall thicknesses and water depths see page 120 and pages 122 through 140.

For unlisted sizes and wall thicknesses extrapolate from above or other average pipelay tables.

# AVERAGE PIPELAY TABLE (Water Depth to 50 Feet)

**Lay or Derrick Barge Fabricate and Install**

NUMBER OF SINGLE RANDOM JOINTS OF CONCRETE WEIGHT COATED PIPE PER TWENTY-FOUR (24) HOUR WORKING DAY

| O.D. Pipe Size Inches | PIPE WALL THICKNESS IN INCHES | | | | | |
|---|---|---|---|---|---|---|
| | 0.6875 | 0.750 | 0.8125 | 0.875 | 0.9375 | 1.0000 |
| 20.000 | 313 | 307 | 301 | 301 | 271 | 244 |
| 22.000 | 298 | 293 | 287 | 287 | 258 | 233 |
| 24.000 | 285 | 279 | 274 | 274 | 246 | 222 |
| 26.000 | 277 | 271 | 266 | 266 | 239 | 215 |
| 28.000 | 268 | 262 | 257 | 257 | 231 | 208 |
| 30.000 | 259 | 254 | 249 | 249 | 224 | 202 |
| 32.000 | 250 | 245 | 240 | 240 | 216 | 194 |
| 34.000 | 240 | 235 | 231 | 231 | 208 | 187 |
| 36.000 | 231 | 226 | 223 | 221 | 199 | 179 |
| 40.000 | 214 | 210 | 206 | 205 | 185 | 166 |
| 42.000 | 197 | 193 | 190 | 189 | 170 | 153 |
| 44.000 | 148 | 141 | 134 | 127 | 114 | 103 |
| 48.000 | 111 | 106 | 100 | 95 | 86 | 77 |
| 52.000 | 83 | 79 | 75 | 71 | 64 | 58 |
| 56.000 | 62 | 59 | 56 | 54 | 48 | 43 |
| 60.000 | 46 | 45 | 42 | 40 | 36 | 33 |

The related construction labor and equipment spreads that follow in this section should be ample for installing the above quantities of submerged, concrete weight coated pipe working a twenty-four (24) hour day.

The above single random joints of submerged pipelay are based on assembling concrete weight coated pipe on a lay barge or combination lay/derrick barge properly rigged with equipment necessary for performing the operations involved.

For other sizes, wall thicknesses and water depths not shown see pages 120 and 121 and 123 through 140.

For unlisted sizes and wall thicknesses extrapolate from above or other average pipelay tables.

# AVERAGE PIPELAY TABLE
# (Water Depth from 50 to 75 Feet)

**Lay or Derrick Barge Fabricate and Install**

NUMBER OF SINGLE RANDOM JOINTS OF CONCRETE WEIGHT COATED PIPE PER TWENTY-FOUR (24) HOUR WORKING DAY

| | PIPE WALL THICKNESS IN INCHES | | | | | |
|---|---|---|---|---|---|---|
| O.D. Pipe Size Inches | 0.312 through 0.337 | 0.365 through 0.375 | 0.432 through 0.438 | 0.500 through 0.531 | 0.5625 through 0.625 | 0.675 — — |
| 4.500 | 530 | — | 516 | 506 | — | 491 |
| 5.500 | — | — | — | — | — | 382 |
| 5.563 | — | 489 | 482 | 480 | 469 | — |
| 6.625 | — | 461 | 458 | 452 | 441 | — |
| 8.625 | — | 409 | 405 | 400 | 393 | — |
| 10.750 | — | 402 | 399 | 394 | 387 | — |
| 12.750 | — | 361 | 358 | 354 | 347 | — |
| 14.000 | — | 354 | 352 | 350 | 344 | — |
| 16.000 | — | 337 | 334 | 330 | 324 | — |
| 18.000 | — | 313 | 310 | 308 | 301 | — |
| 20.000 | — | 297 | 294 | 291 | 285 | — |
| 22.000 | — | 282 | 279 | 277 | 271 | — |
| 24.000 | — | 270 | 268 | 265 | 259 | — |

The related construction labor and equipment spreads that follow in this section should be ample for installing the above quantities of submerged, concrete weight coated pipe working a twenty-four (24) hour day.

The above single random joints of submerged pipelay are based on assembling concrete weight coated pipe on a lay barge or combination lay/derrick barge properly rigged with equipment necessary for performing the operations involved.

For other sizes, wall thicknesses and water depths not shown, see pages 120 through 140.

For unlisted sizes and wall thicknesses extrapolate from above or other average pipelay tables.

If double random lengths of pipe are to be installed, use 65 percent of the above number of joints.

# AVERAGE PIPELAY TABLE
# (Water Depth from 50 to 75 Feet)

**Lay or Derrick Barge Fabricate and Install**

NUMBER OF SINGLE RANDOM JOINTS OF CONCRETE WEIGHT COATED PIPE PER TWENTY-FOUR (24) HOUR WORKING DAY

| O.D. Pipe Size Inches | PIPE WALL THICKNESS IN INCHES | | | | |
|---|---|---|---|---|---|
| | 0.375 | 0.4325 | 0.500 | 0.5625 | 0.625 |
| 26.000 | 262 | 259 | 257 | 254 | 252 |
| 28.000 | 254 | 251 | 248 | 247 | 244 |
| 30.000 | 246 | 243 | 241 | 238 | 236 |
| 32.000 | 237 | 234 | 232 | 229 | 228 |
| 34.000 | 228 | 225 | 223 | 220 | 218 |
| 36.000 | 218 | 216 | 214 | 212 | 210 |
| 40.000 | — | 201 | 198 | 197 | 195 |
| 42.000 | — | 186 | 183 | 182 | 179 |
| 44.000 | — | 139 | 137 | 137 | 135 |
| 48.000 | — | 105 | 103 | 102 | 101 |
| 52.000 | — | 78 | 77 | 76 | 75 |
| 56.000 | — | 58 | 57 | 57 | 56 |
| 60.000 | — | 44 | 43 | 43 | 42 |

The related construction labor and equipment spreads that follow in this section should be ample for installing the above quantities of submerged, concrete weight coated pipe working a twenty-four (24) hour day.

The above single random joints of submerged pipelay are based on assembling concrete weight coated pipe on a lay barge or combination lay/derrick barge properly rigged with equipment necessary for performing the operations involved.

For other sizes, wall thicknesses and water depths not shown, see pages 120 through 140.

For unlisted sizes and wall thicknesses extrapolate from above or other average pipelay tables.

# AVERAGE PIPELAY TABLE (Water Depth from 50 to 75 Feet)

**Lay or Derrick Barge Fabricate and Install**

NUMBER OF SINGLE RANDOM JOINTS OF CONCRETE WEIGHT COATED PIPE PER TWENTY-FOUR (24) HOUR WORKING DAY

| O.D. Pipe Size Inches | PIPE WALL THICKNESS IN INCHES | | | | | |
|---|---|---|---|---|---|---|
| | 0.6875 | 0.750 | 0.8125 | 0.875 | 0.9375 | 1.0000 |
| 20.000 | 285 | 279 | 274 | 274 | 247 | 222 |
| 22.000 | 271 | 267 | 261 | 261 | 235 | 212 |
| 24.000 | 259 | 254 | 249 | 249 | 224 | 202 |
| 26.000 | 252 | 247 | 242 | 242 | 217 | 196 |
| 28.000 | 244 | 238 | 234 | 234 | 210 | 189 |
| 30.000 | 236 | 231 | 227 | 227 | 204 | 184 |
| 32.000 | 228 | 223 | 218 | 218 | 197 | 177 |
| 34.000 | 218 | 214 | 210 | 210 | 189 | 170 |
| 36.000 | 210 | 206 | 203 | 201 | 181 | 163 |
| 40.000 | 195 | 191 | 187 | 187 | 168 | 151 |
| 42.000 | 179 | 176 | 173 | 172 | 155 | 139 |
| 44.000 | 135 | 128 | 122 | 116 | 104 | 94 |
| 48.000 | 101 | 96 | 91 | 86 | 78 | 70 |
| 52.000 | 75 | 72 | 68 | 65 | 58 | 53 |
| 56.000 | 56 | 54 | 51 | 49 | 44 | 39 |
| 60.000 | 42 | 41 | 38 | 36 | 33 | 30 |

The related construction labor and equipment spreads that follow in this section should be ample for installing the above quantities of submerged, concrete weight coated pipe working a twenty-four (24) hour day.

The above single random joints of submerged pipelay are based on assembling concrete weight coated pipe on a lay barge or combination lay/derrick barge properly rigged with equipment necessary for performing the operations involved.

For other sizes, wall thicknesses and water depths not shown, see pages 120 through 140.

For unlisted sizes and wall thicknesses extrapolate from above or other average pipelay tables.

# AVERAGE PIPELAY TABLE
# (Water Depth from 75 to 100 Feet)

**Lay or Derrick Barge Fabricate and Install**

NUMBER OF SINGLE RANDOM JOINTS OF CONCRETE WEIGHT COATED PIPE PER TWENTY-FOUR (24) HOUR WORKING DAY

| | PIPE WALL THICKNESS IN INCHES | | | | | |
|---|---|---|---|---|---|---|
| O.D. Pipe Size Inches | 0.312 through 0.337 | 0.365 through 0.375 | 0.432 through 0.438 | 0.500 through 0.531 | 0.5625 through 0.625 | 0.675 — — |
| 4.500 | 466 | — | 454 | 445 | — | 432 |
| 5.500 | — | — | — | — | — | 336 |
| 5.563 | — | 430 | 424 | 422 | 412 | — |
| 6.625 | — | 406 | 402 | 398 | 388 | — |
| 8.625 | — | 359 | 356 | 352 | 346 | — |
| 10.750 | — | 354 | 350 | 346 | 340 | — |
| 12.750 | — | 318 | 314 | 311 | 305 | — |
| 14.000 | — | 311 | 310 | 308 | 302 | — |
| 16.000 | — | 296 | 294 | 290 | 285 | — |
| 18.000 | — | 275 | 273 | 270 | 265 | — |
| 20.000 | — | 261 | 258 | 256 | 250 | — |
| 22.000 | — | 248 | 246 | 243 | 238 | — |
| 24.000 | — | 238 | 235 | 233 | 228 | — |

The related construction labor and equipment spreads that follow in this section should be ample for installing the above quantities of submerged, concrete weight coated pipe working a twenty-four (24) hour day.

The above single random joints of submerged pipelay are based on assembling concrete weight coated pipe on a lay barge or combination lay/derrick barge properly rigged with equipment necessary for performing the operations involved.

For other sizes, wall thicknesses and water depths not shown, see pages 120 through 140.

For unlisted sizes and wall thicknesses extrapolate from above or other average pipelay tables.

If double random lengths of pipe are to be installed, use 65 percent of the above number of joints.

# AVERAGE PIPELAY TABLE
# (Water Depth from 75 to 100 Feet)

**Lay or Derrick Barge Fabricate and Install**

NUMBER OF SINGLE RANDOM JOINTS OF CONCRETE WEIGHT COATED PIPE PER TWENTY-FOUR (24) HOUR WORKING DAY

| O.D. Pipe Size Inches | PIPE WALL THICKNESS IN INCHES | | | | |
|---|---|---|---|---|---|
| | 0.375 | 0.4325 | 0.500 | 0.5625 | 0.625 |
| 26.000 | 230 | 228 | 226 | 223 | 222 |
| 28.000 | 223 | 221 | 218 | 217 | 214 |
| 30.000 | 216 | 214 | 212 | 210 | 207 |
| 32.000 | 208 | 206 | 204 | 202 | 200 |
| 34.000 | 200 | 198 | 196 | 194 | 192 |
| 36.000 | 192 | 190 | 188 | 186 | 185 |
| 40.000 | — | 177 | 174 | 173 | 171 |
| 42.000 | — | 163 | 161 | 160 | 158 |
| 44.000 | — | 122 | 121 | 120 | 118 |
| 48.000 | — | 92 | 90 | 89 | 88 |
| 52.000 | — | 69 | 68 | 67 | 66 |
| 56.000 | — | 51 | 50 | — | — |
| 60.000 | — | 38 | 37 | — | — |

The related construction labor and equipment spreads that follow in this section should be ample for installing the above quantities of submerged, concrete weight coated pipe working a twenty-four (24) hour day.

The above single random joints of submerged pipelay are based on assembling concrete weight coated pipe on a lay barge or combination lay/derrick barge properly rigged with equipment necessary for performing the operations involved.

For other sizes, wall thicknesses and water depths not shown, see pages 120 through 140.

For unlisted sizes and wall thicknesses extrapolate from above or other average pipelay tables.

# AVERAGE PIPELAY TABLE (Water Depth from 75 to 100 Feet)

**Lay or Derrick Barge Fabricate and Install**

NUMBER OF SINGLE RANDOM JOINTS OF CONCRETE WEIGHT COATED PIPE PER TWENTY-FOUR (24) HOUR WORKING DAY

| O.D. Pipe Size Inches | PIPE WALL THICKNESS IN INCHES | | | | | |
|---|---|---|---|---|---|---|
| | 0.6875 | 0.750 | 0.8125 | 0.875 | 0.9375 | 1.0000 |
| 20.000 | 250 | 246 | 241 | 241 | 217 | 195 |
| 22.000 | 238 | 234 | 230 | 230 | 206 | 186 |
| 24.000 | 228 | 223 | 219 | 219 | 197 | 178 |
| 26.000 | 222 | 217 | 213 | 213 | 191 | 172 |
| 28.000 | 214 | 210 | 206 | 206 | 185 | 166 |
| 30.000 | 207 | 203 | 199 | 199 | 179 | 162 |
| 32.000 | 200 | 196 | 192 | 192 | 173 | 155 |
| 34.000 | 192 | 188 | 185 | 185 | 166 | 150 |
| 36.000 | 185 | 181 | 178 | 177 | 159 | 143 |
| 40.000 | 171 | 168 | 165 | 164 | 148 | 133 |
| 42.000 | 158 | 154 | 152 | 151 | 136 | 122 |
| 44.000 | 118 | 113 | 107 | 102 | 91 | 82 |
| 48.000 | 89 | 85 | 80 | 76 | 69 | 62 |
| 52.000 | 66 | 63 | 60 | 57 | 51 | 46 |
| 56.000 | — | — | — | — | — | — |
| 60.000 | — | — | — | — | — | — |

The related construction labor and equipment spreads that follow in this section should be ample for installing the above quantities of submerged, concrete weight coated pipe working a twenty-four (24) hour day.

The above single random joints of submerged pipelay are based on assembling concrete weight coated pipe on a lay barge or combination lay/derrick barge properly rigged with equipment necessary for performing the operations involved.

For other sizes, wall thicknesses and water depths not shown, see pages 120 through 140.

For unlisted sizes and wall thicknesses extrapolate from above or other average pipelay tables.

# AVERAGE PIPELAY TABLE (Water Depth from 100 to 125 Feet)

**Lay or Derrick Barge Fabricate and Install**

NUMBER OF SINGLE RANDOM JOINTS OF CONCRETE WEIGHT COATED PIPE PER TWENTY-FOUR (24) HOUR WORKING DAY

| | PIPE WALL THICKNESS IN INCHES | | | | | |
|---|---|---|---|---|---|---|
| O.D. Pipe Size Inches | 0.312 through 0.337 | 0.365 through 0.375 | 0.432 through 0.438 | 0.500 through 0.531 | 0.5625 through 0.625 | 0.675 — — |
| 4.500 | 443 | — | 431 | 423 | — | 410 |
| 5.500 | — | — | — | — | — | 319 |
| 5.563 | — | 408 | 403 | 401 | 391 | — |
| 6.625 | — | 385 | 382 | 378 | 369 | — |
| 8.625 | — | 341 | 338 | 334 | 328 | — |
| 10.750 | — | 336 | 333 | 329 | 323 | — |
| 12.750 | — | 302 | 299 | 296 | 290 | — |
| 14.000 | — | 296 | 294 | 293 | 287 | — |
| 16.000 | — | 281 | 279 | 276 | 271 | — |
| 18.000 | — | 261 | 259 | 257 | 252 | — |
| 20.000 | — | 248 | 245 | 243 | 238 | — |
| 22.000 | — | 236 | 233 | 231 | 226 | — |
| 24.000 | — | 226 | 223 | 221 | 217 | — |

The related construction labor and equipment spreads that follow in this section should be ample for installing the above quantities of submerged, concrete weight coated pipe working a twenty-four (24) hour day.

The above single random joints of submerged pipelay are based on assembling concrete weight coated pipe on a lay barge or combination lay/derrick barge properly rigged with equipment necessary for performing the operations involved.

For other sizes, wall thicknesses and water depths not shown, see pages 120 through 140.

For unlisted sizes and wall thicknesses extrapolate from above or other average pipelay tables.

If double random lengths of pipe are to be installed, use 65 percent of the above number of joints.

# AVERAGE PIPELAY TABLE
# (Water Depth from 100 to 125 Feet)

**Lay or Derrick Barge Fabricate and Install**

NUMBER OF SINGLE RANDOM JOINTS OF CONCRETE WEIGHT COATED PIPE PER TWENTY-FOUR (24) HOUR WORKING DAY

| O.D. Pipe Size Inches | PIPE WALL THICKNESS IN INCHES | | | | |
|---|---|---|---|---|---|
| | 0.375 | 0.4325 | 0.500 | 0.5625 | 0.625 |
| 26.000 | 219 | 217 | 214 | 212 | 211 |
| 28.000 | 212 | 210 | 207 | 206 | 204 |
| 30.000 | 205 | 203 | 201 | 199 | 197 |
| 32.000 | 198 | 195 | 194 | 192 | 190 |
| 34.000 | 190 | 188 | 186 | 184 | 182 |
| 36.000 | 182 | 180 | 179 | 177 | 176 |
| 40.000 | — | 168 | 166 | 164 | 163 |
| 42.000 | — | 155 | 153 | 152 | 150 |
| 44.000 | — | 116 | 115 | 114 | 112 |
| 48.000 | — | 87 | 86 | 85 | 84 |
| 52.000 | — | 65 | 64 | 63 | 62 |
| 56.000 | — | — | — | — | — |
| 60.000 | — | — | — | — | — |

The related construction labor and equipment spreads that follow in this section should be ample for installing the above quantities of submerged, concrete weight coated pipe working a twenty-four (24) hour day.

The above single random joints of submerged pipelay are based on assembling concrete weight coated pipe on a lay barge or combination lay/derrick barge properly rigged with equipment necessary for performing the operations involved.

For other sizes, wall thicknesses and water depths not shown, see pages 120 through 140.

For unlisted sizes and wall thicknesses extrapolate from above or other average pipelay tables.

# AVERAGE PIPELAY TABLE
# (Water Depth from 100 to 125 Feet)

**Lay or Derrick Barge Fabricate and Install**

NUMBER OF SINGLE RANDOM JOINTS OF CONCRETE WEIGHT COATED PIPE PER TWENTY-FOUR (24) HOUR WORKING DAY

| O.D. Pipe Size Inches | PIPE WALL THICKNESS IN INCHES | | | | | |
|---|---|---|---|---|---|---|
| | 0.6875 | 0.750 | 0.8125 | 0.875 | 0.9375 | 1.0000 |
| 20.000 | 238 | 233 | 229 | 229 | 206 | 185 |
| 22.000 | 226 | 223 | 218 | 218 | 196 | 177 |
| 24.000 | 217 | 212 | 208 | 208 | 187 | 169 |
| 26.000 | 211 | 206 | 202 | 202 | 182 | 163 |
| 28.000 | 204 | 199 | 195 | 195 | 176 | 158 |
| 30.000 | 195 | 193 | 189 | 189 | 170 | 154 |
| 32.000 | 190 | 186 | 182 | 182 | 164 | 147 |
| 34.000 | 182 | 179 | 176 | 176 | 158 | 142 |
| 36.000 | 176 | 172 | 169 | 168 | 151 | 136 |
| 40.000 | 163 | 160 | 157 | 156 | 141 | 126 |
| 42.000 | 150 | 147 | 144 | 143 | 129 | 116 |
| 44.000 | 112 | 107 | 102 | 97 | 87 | 78 |
| 48.000 | 84 | 81 | 76 | 72 | 65 | 59 |
| 52.000 | 63 | 60 | 57 | 54 | 49 | 44 |
| 56.000 | — | — | — | — | — | — |
| 60.000 | — | — | — | — | — | — |

The related construction labor and equipment spreads that follow in this section should be ample for installing the above quantities of submerged, concrete weight coated pipe working a twenty-four (24) hour day.

The above single random joints of submerged pipelay are based on assembling concrete weight coated pipe on a lay barge or combination lay/derrick barge properly rigged with equipment necessary for performing the operations involved.

For other sizes, wall thicknesses and water depths not shown, see pages 120 through 140.

For unlisted sizes and wall thicknesses extrapolate from above or other average pipelay tables.

# AVERAGE PIPELAY TABLE
# (Water Depth from 125 to 150 Feet)

**Lay or Derrick Barge Fabricate and Install**

NUMBER OF SINGLE RANDOM JOINTS OF CONCRETE WEIGHT COATED PIPE PER TWENTY-FOUR (24) HOUR WORKING DAY

| O.D. Pipe Size Inches | PIPE WALL THICKNESS IN INCHES | | | | | |
|---|---|---|---|---|---|---|
| | 0.312 through 0.337 | 0.365 through 0.375 | 0.432 through 0.438 | 0.500 through 0.531 | 0.5625 through 0.625 | 0.675 — — |
| 4.500 | 408 | — | 397 | 389 | — | 378 |
| 5.500 | — | — | — | — | — | 294 |
| 5.563 | — | 376 | 371 | 369 | 361 | — |
| 6.625 | — | 355 | 352 | 348 | 340 | — |
| 8.625 | — | 314 | 312 | 308 | 302 | — |
| 10.750 | — | 309 | 307 | 303 | 298 | — |
| 12.750 | — | 278 | 275 | 272 | 267 | — |
| 14.000 | — | 272 | 271 | 270 | 265 | — |
| 16.000 | — | 259 | 257 | 254 | 249 | — |
| 18.000 | — | 241 | 239 | 237 | 232 | — |
| 20.000 | — | 228 | 226 | 224 | 219 | — |
| 22.000 | — | 217 | 215 | 213 | 209 | — |
| 24.000 | — | 208 | 206 | 204 | 200 | — |

The related construction labor and equipment spreads that follow in this section should be ample for installing the above quantities of submerged, concrete weight coated pipe working a twenty-four (24) hour day.

The above single random joints of submerged pipelay are based on assembling concrete weight coated pipe on a lay barge or combination lay/derrick barge properly rigged with equipment necessary for performing the operations involved.

For other sizes, wall thicknesses and water depths not shown, see pages 120 through 140.

For unlisted sizes and wall thicknesses extrapolate from above or other average pipelay tables.

If double random lengths of pipe are to be installed, use 65 percent of the above number of joints.

# AVERAGE PIPELAY TABLE
# (Water Depth from 125 to 150 Feet)

**Lay or Derrick Barge Fabricate and Install**

NUMBER OF SINGLE RANDOM JOINTS OF CONCRETE WEIGHT COATED PIPE PER TWENTY-FOUR (24) HOUR WORKING DAY

| O.D. Pipe Size Inches | PIPE WALL THICKNESS IN INCHES | | | | |
|---|---|---|---|---|---|
| | 0.375 | 0.4325 | 0.500 | 0.5625 | 0.625 |
| 26.000 | 202 | 200 | 197 | 195 | 194 |
| 28.000 | 195 | 193 | 191 | 190 | 188 |
| 30.000 | 189 | 187 | 186 | 183 | 181 |
| 32.000 | 182 | 180 | 179 | 176 | 175 |
| 34.000 | 175 | 173 | 172 | 169 | 168 |
| 36.000 | 168 | 166 | 165 | 163 | 162 |
| 40.000 | — | 155 | 153 | 151 | 150 |
| 42.000 | — | 143 | 141 | 140 | 138 |
| 44.000 | — | 107 | 106 | 105 | 104 |
| 48.000 | — | 81 | 79 | 78 | 77 |
| 52.000 | — | 60 | — | — | — |
| 56.000 | — | — | — | — | — |
| 60.000 | — | — | — | — | — |

The related construction labor and equipment spreads that follow in this section should be ample for installing the above quantities of submerged, concrete weight coated pipe working a twenty-four (24) hour day.

The above single random joints of submerged pipelay are based on assembling concrete weight coated pipe on a lay barge or combination lay/derrick barge properly rigged with equipment necessary for performing the operations involved.

For other sizes, wall thicknesses and water depths not shown, see pages 120 through 140.

For unlisted sizes and wall thicknesses extrapolate from above or other average pipelay tables.

# AVERAGE PIPELAY TABLE (Water Depth from 125 to 150 Feet)

**Lay or Derrick Barge Fabricate and Install**

NUMBER OF SINGLE RANDOM JOINTS OF CONCRETE WEIGHT COATED PIPE PER TWENTY-FOUR (24) HOUR WORKING DAY

| O.D. Pipe Size Inches | PIPE WALL THICKNESS IN INCHES | | | | | |
|---|---|---|---|---|---|---|
| | 0.6875 | 0.750 | 0.8125 | 0.875 | 0.9375 | 1.0000 |
| 20.000 | 219 | 215 | 211 | 211 | 190 | 171 |
| 22.000 | 209 | 205 | 201 | 201 | 181 | 163 |
| 24.000 | 200 | 195 | 192 | 192 | 172 | 155 |
| 26.000 | 194 | 190 | 186 | 186 | 167 | 151 |
| 28.000 | 188 | 183 | 180 | 180 | 162 | 146 |
| 30.000 | 181 | 178 | 174 | 174 | 157 | 141 |
| 32.000 | 175 | 172 | 168 | 168 | 151 | 136 |
| 34.000 | 168 | 165 | 162 | 162 | 146 | 131 |
| 36.000 | 162 | 158 | 156 | 155 | 139 | 125 |
| 40.000 | 150 | 147 | 144 | 143 | 130 | 116 |
| 42.000 | 138 | 135 | 133 | 132 | 119 | 107 |
| 44.000 | 104 | 99 | 94 | 89 | 80 | 72 |
| 48.000 | 78 | 74 | 70 | 67 | 60 | 54 |
| 52.000 | — | — | — | — | — | — |
| 56.000 | — | — | — | — | — | — |
| 60.000 | — | — | — | — | — | — |

The related construction labor and equipment spreads that follow in this section should be ample for installing the above quantities of submerged, concrete weight coated pipe working a twenty-four (24) hour day.

The above single random joints of submerged pipelay are based on assembling concrete weight coated pipe on a lay barge or combination lay/derrick barge properly rigged with equipment necessary for performing the operations involved.

For other sizes, wall thicknesses and water depths not shown, see pages 120 through 140.

For unlisted sizes and wall thicknesses extrapolate from above or other average pipelay tables.

# AVERAGE PIPELAY TABLE
# (Water Depth from 150 to 200 Feet)

**Lay or Derrick Barge Fabricate and Install**

NUMBER OF SINGLE RANDOM JOINTS OF CONCRETE WEIGHT COATED PIPE PER TWENTY-FOUR (24) HOUR WORKING DAY

| | PIPE WALL THICKNESS IN INCHES | | | | | |
|---|---|---|---|---|---|---|
| O.D. Pipe Size Inches | 0.312 through 0.337 | 0.365 through 0.375 | 0.432 through 0.438 | 0.500 through 0.531 | 0.5625 through 0.625 | 0.675 — — |
| 4.500 | 367 | — | 357 | 350 | — | 340 |
| 5.500 | — | — | — | — | — | 265 |
| 5.563 | — | 338 | 334 | 332 | 324 | — |
| 6.625 | — | 319 | 317 | 313 | 306 | — |
| 8.625 | — | 283 | 280 | 277 | 272 | — |
| 10.750 | — | 278 | 276 | 273 | 268 | — |
| 12.750 | — | 250 | 248 | 245 | 240 | — |
| 14.000 | — | 245 | 244 | 243 | 238 | — |
| 16.000 | — | 233 | 231 | 229 | 224 | — |
| 18.000 | — | 217 | 215 | 213 | 209 | — |
| 20.000 | — | 205 | 203 | 202 | 197 | — |
| 22.000 | — | 195 | 193 | 192 | 188 | — |
| 24.000 | — | 187 | 185 | 183 | 180 | — |

The related construction labor and equipment spreads that follow in this section should be ample for installing the above quantities of submerged, concrete weight coated pipe working a twenty-four (24) hour day.

The above single random joints of submerged pipelay are based on assembling concrete weight coated pipe on a lay barge or combination lay/derrick barge properly rigged with equipment necessary for performing the operations involved.

For other sizes, wall thicknesses and water depths not shown, see pages 120 through 140.

For unlisted sizes and wall thicknesses extrapolate from above or other average pipelay tables.

If double random lengths of pipe are to be installed, use 65 percent of the above number of joints.

# AVERAGE PIPELAY TABLE
# (Water Depth from 150 to 200 Feet)

**Lay or Derrick Barge Fabricate and Install**

NUMBER OF SINGLE RANDOM JOINTS OF CONCRETE WEIGHT COATED PIPE PER TWENTY-FOUR (24) HOUR WORKING DAY

| O.D. Pipe Size Inches | PIPE WALL THICKNESS IN INCHES | | | | |
|---|---|---|---|---|---|
| | 0.375 | 0.4325 | 0.500 | 0.5625 | 0.625 |
| 26.000 | 181 | 180 | 178 | 176 | 175 |
| 28.000 | 176 | 174 | 172 | 171 | 169 |
| 30.000 | 170 | 168 | 167 | 165 | 163 |
| 32.000 | 164 | 162 | 161 | 159 | 158 |
| 34.000 | 158 | 156 | 154 | 152 | 151 |
| 36.000 | 151 | 149 | 148 | 147 | 146 |
| 40.000 | — | 139 | 137 | 136 | 135 |
| 42.000 | — | 129 | 127 | 126 | 124 |
| 44.000 | — | 96 | 95 | 94 | 93 |
| 48.000 | — | 72 | 71 | — | — |
| 52.000 | — | — | — | — | — |
| 56.000 | — | — | — | — | — |
| 60.000 | — | — | — | — | — |

The related construction labor and equipment spreads that follow in this section should be ample for installing the above quantities of submerged, concrete weight coated pipe working a twenty-four (24) hour day.

The above single random joints of submerged pipelay are based on assembling concrete weight coated pipe on a lay barge or combination lay/derrick barge properly rigged with equipment necessary for performing the operations involved.

For other sizes, wall thicknesses and water depths not shown, see pages 120 through 140.

For unlisted sizes and wall thicknesses extrapolate from above or other average pipelay tables.

# AVERAGE PIPELAY TABLE
# (Water Depth from 150 to 200 Feet)

**Lay or Derrick Barge Fabricate and Install**

NUMBER OF SINGLE RANDOM JOINTS OF CONCRETE WEIGHT COATED PIPE PER TWENTY-FOUR (24) HOUR WORKING DAY

| O.D. Pipe Size Inches | PIPE WALL THICKNESS IN INCHES | | | | | |
|---|---|---|---|---|---|---|
| | 0.6875 | 0.750 | 0.8125 | 0.875 | 0.9375 | 1.0000 |
| 20.000 | 197 | 193 | 190 | 190 | 171 | 154 |
| 22.000 | 188 | 185 | 181 | 181 | 163 | 147 |
| 24.000 | 180 | 176 | 173 | 173 | 155 | 140 |
| 26.000 | 175 | 171 | 168 | 168 | 151 | 135 |
| 28.000 | 169 | 165 | 162 | 162 | 146 | 131 |
| 30.000 | 163 | 160 | 157 | 157 | 141 | 127 |
| 32.000 | 158 | 154 | 151 | 151 | 136 | 122 |
| 34.000 | 151 | 148 | 146 | 146 | 131 | 118 |
| 36.000 | 146 | 142 | 140 | 139 | 125 | 113 |
| 40.000 | 135 | 132 | 130 | 129 | 117 | 105 |
| 42.000 | 124 | 122 | 120 | 119 | 107 | 96 |
| 44.000 | 93 | 89 | 84 | 80 | 72 | 65 |
| 48.000 | — | — | — | — | — | — |
| 52.000 | — | — | — | — | — | — |
| 56.000 | — | — | — | — | — | — |
| 60.000 | — | — | — | — | — | — |

The related construction labor and equipment spreads that follow in this section should be ample for installing the above quantities of submerged, concrete weight coated pipe working a twenty-four (24) hour day.

The above single random joints of submerged pipelay are based on assembling concrete weight coated pipe on a lay barge or combination lay/derrick barge properly rigged with equipment necessary for performing the operations involved.

For other sizes, wall thicknesses and water depths not shown, see pages 120 through 140.

For unlisted sizes and wall thicknesses extrapolate from above or other average pipelay tables.

# AVERAGE PIPELAY TABLE
# (Water Depth from 200 to 250 Feet)

**Lay or Derrick Barge Fabricate and Install**

NUMBER OF SINGLE RANDOM JOINTS OF CONCRETE WEIGHT COATED PIPE PER TWENTY-FOUR (24) HOUR WORKING DAY

| O.D. Pipe Size Inches | PIPE WALL THICKNESS IN INCHES | | | | | |
|---|---|---|---|---|---|---|
| | 0.312 through 0.337 | 0.365 through 0.375 | 0.432 through 0.438 | 0.500 through 0.531 | 0.5625 through 0.625 | 0.675 — — |
| 4.500 | 326 | — | 318 | 311 | — | 302 |
| 5.500 | — | — | — | — | — | 235 |
| 5.563 | — | 301 | 297 | 295 | 288 | — |
| 6.625 | — | 284 | 282 | 278 | 272 | — |
| 8.625 | — | 251 | 249 | 246 | 242 | — |
| 10.750 | — | 248 | 245 | 242 | 238 | — |
| 12.750 | — | 222 | 220 | 218 | 213 | — |
| 14.000 | — | 218 | 217 | 216 | 212 | — |
| 16.000 | — | 207 | 206 | 203 | 199 | — |
| 18.000 | — | 193 | 191 | 189 | 185 | — |
| 20.000 | — | 183 | 181 | 179 | 175 | — |
| 22.000 | — | 174 | 172 | 170 | 167 | — |
| 24.000 | — | 166 | 165 | 163 | 160 | — |

The related construction labor and equipment spreads that follow in this section should be ample for installing the above quantities of submerged, concrete weight coated pipe working a twenty-four (24) hour day.

The above single random joints of submerged pipelay are based on assembling concrete weight coated pipe on a lay barge or combination lay/derrick barge properly rigged with equipment necessary for performing the operations involved.

For other sizes, wall thicknesses and water depths not shown, see pages 120 through 140.

For unlisted sizes and wall thicknesses extrapolate from above or other average pipelay tables.

If double random lengths of pipe are to be installed, use 65 percent of the above number of joints.

# AVERAGE PIPELAY TABLE (Water Depth from 200 to 250 Feet)

### Lay or Derrick Barge Fabricate and Install

NUMBER OF SINGLE RANDOM JOINTS OF CONCRETE WEIGHT COATED PIPE PER TWENTY-FOUR (24) HOUR WORKING DAY

| O.D. Pipe Size Inches | PIPE WALL THICKNESS IN INCHES | | | | |
|---|---|---|---|---|---|
| | 0.375 | 0.4325 | 0.500 | 0.5625 | 0.625 |
| 26.000 | 161 | 160 | 158 | 156 | 155 |
| 28.000 | 156 | 155 | 153 | 152 | 150 |
| 30.000 | 151 | 150 | 148 | 147 | 145 |
| 32.000 | 146 | 144 | 143 | 141 | 140 |
| 34.000 | 140 | 138 | 137 | 136 | 134 |
| 36.000 | 134 | 133 | 132 | 130 | 129 |
| 40.000 | — | 124 | 122 | 121 | 120 |
| 42.000 | — | 114 | 113 | 112 | 110 |
| 44.000 | — | 86 | 85 | 84 | 83 |
| 48.000 | — | 64 | 63 | — | — |
| 52.000 | — | — | — | — | — |
| 56.000 | — | — | — | — | — |
| 60.000 | — | — | — | — | — |

The related construction labor and equipment spreads that follow in this section should be ample for installing the above quantities of submerged, concrete weight coated pipe working a twenty-four (24) hour day.

The above single random joints of submerged pipelay are based on assembling concrete weight coated pipe on a lay barge or combination lay/derrick barge properly rigged with equipment necessary for performing the operations involved.

For other sizes, wall thicknesses and water depths not shown, see pages 120 through 140.

For unlisted sizes and wall thicknesses extrapolate from above or other average pipelay tables.

If double random lengths of pipe are to be installed, use 65 percent of the above number of joints.

# AVERAGE PIPELAY TABLE
# (Water Depth from 200 to 250 Feet)

**Lay or Derrick Barge Fabricate and Install**

NUMBER OF SINGLE RANDOM JOINTS OF CONCRETE WEIGHT COATED PIPE PER TWENTY-FOUR (24) HOUR WORKING DAY

| O.D. Pipe Size Inches | PIPE WALL THICKNESS IN INCHES | | | | | |
|---|---|---|---|---|---|---|
| | 0.6875 | 0.750 | 0.8125 | 0.875 | 0.9375 | 1.0000 |
| 20.000 | 175 | 172 | 169 | 169 | 152 | 137 |
| 22.000 | 167 | 164 | 161 | 161 | 144 | 130 |
| 24.000 | 160 | 156 | 153 | 153 | 138 | 124 |
| 26.000 | 155 | 152 | 149 | 149 | 134 | 120 |
| 28.000 | 150 | 147 | 144 | 144 | 129 | 116 |
| 30.000 | 145 | 142 | 139 | 139 | 125 | 113 |
| 32.000 | 140 | 137 | 134 | 134 | 121 | 109 |
| 34.000 | 134 | 132 | 129 | 129 | 116 | 105 |
| 36.000 | 129 | 127 | 125 | 124 | 111 | 100 |
| 40.000 | 120 | 118 | 115 | 114 | 104 | 93 |
| 42.000 | 110 | 108 | 106 | 105 | 95 | 86 |
| 44.000 | — | — | — | — | — | — |
| 48.000 | — | — | — | — | — | — |
| 52.000 | — | — | — | — | — | — |
| 56.000 | — | — | — | — | — | — |
| 60.000 | — | — | — | — | — | — |

The related construction labor and equipment spreads that follow in this section should be ample for installing the above quantities of submerged, concrete weight coated pipe working a twenty-four (24) hour day.

The above single random joints of submerged pipelay are based on assembling concrete weight coated pipe on a lay barge or combination lay/derrick barge properly rigged with equipment necessary for performing the operations involved.

For other sizes, wall thicknesses and water depths not shown, see pages 120 through 139.

For unlisted sizes and wall thicknesses extrapolate from above or other average pipelay tables.

# AVERAGE PIPELINE TO PLATFORM RISER INSTALLATION (Water Depth to 150 Feet)

CREW AND EQUIPMENT DAYS PER RISER

| O.D. Pipe Size Inches | PIPE WALL THICKNESS | | | | | | | | | |
|---|---|---|---|---|---|---|---|---|---|---|
| | to 0.500 | | | | | 0.500 to 1.000 | | | | |
| | Item Numbers | | | | | Item Numbers | | | | |
| | 1 | 2 | 3 | 4 | T | 1 | 2 | 3 | 4 | T |
| 4.500 | 0.13 | 0.04 | 0.15 | 0.21 | 0.53 | 0.23 | 0.04 | 0.26 | 0.21 | 0.74 |
| 5.500 | 0.15 | 0.04 | 0.17 | 0.21 | 0.57 | 0.25 | 0.04 | 0.29 | 0.21 | 0.79 |
| 5.563 | 0.16 | 0.04 | 0.18 | 0.21 | 0.59 | 0.27 | 0.04 | 0.31 | 0.21 | 0.83 |
| 6.625 | 0.17 | 0.04 | 0.20 | 0.21 | 0.62 | 0.33 | 0.04 | 0.38 | 0.21 | 0.96 |
| 8.625 | 0.19 | 0.04 | 0.24 | 0.25 | 0.72 | 0.42 | 0.04 | 0.49 | 0.25 | 1.20 |
| 10.750 | 0.23 | 0.04 | 0.30 | 0.25 | 0.82 | 0.54 | 0.04 | 0.64 | 0.25 | 1.47 |
| 12.750 | 0.27 | 0.04 | 0.36 | 0.25 | 0.92 | 0.58 | 0.04 | 0.71 | 0.25 | 1.58 |
| 14.000 | 0.33 | 0.04 | 0.43 | 0.33 | 1.13 | 0.76 | 0.04 | 0.90 | 0.33 | 2.03 |
| 16.000 | 0.36 | 0.04 | 0.48 | 0.33 | 1.21 | 0.79 | 0.04 | 0.96 | 0.33 | 2.12 |
| 18.000 | 0.40 | 0.04 | 0.56 | 0.33 | 1.33 | 0.84 | 0.04 | 1.05 | 0.33 | 2.26 |
| 20.000 | 0.43 | 0.08 | 0.60 | 0.38 | 1.49 | 0.86 | 0.08 | 1.10 | 0.38 | 2.42 |

The related construction labor and equipment spreads that follow should be capable of installing the following pipeline to platform pipe risers within the time frames listed above.

Above item numbers represent:

1) Capping off of pipeline and laying on bottom.
2) Measuring on bottom with divers to obtain exact cut-off point for riser connection.
3) Raise pipeline to lay barge, cut off and connect prefabricated riser.
4) Coat and wrap riser where required, lower back to bottom and fasten riser to jacket and platform.

T) Total of the above operations time.

Above time frames do not include fabrication of risers.

If riser is to be installed by platform crew, see related tables under section 7, "Offshore Installation of Equipment and Appurtenances."

# AVERAGE PIPELINE TO PLATFORM RISER INSTALLATION (Water Depth to 150 Feet)

CREW AND EQUIPMENT DAYS PER RISER

| O.D. Pipe Size Inches | PIPE WALL THICKNESS | | | | | | | | | |
|---|---|---|---|---|---|---|---|---|---|---|
| | to 0.500 | | | | | 0.500 to 1.000 | | | | |
| | Item Numbers | | | | | Item Numbers | | | | |
| | 1 | 2 | 3 | 4 | T | 1 | 2 | 3 | 4 | T |
| 22 | 0.47 | 0.08 | 0.68 | 0.38 | 1.61 | 0.89 | 0.08 | 1.18 | 0.38 | 2.53 |
| 24 | 0.52 | 0.08 | 0.77 | 0.38 | 1.75 | 0.91 | 0.08 | 1.27 | 0.38 | 2.64 |
| 26 | 0.61 | 0.08 | 1.02 | 0.46 | 2.17 | 0.98 | 0.08 | 1.47 | 0.46 | 2.99 |
| 28 | 0.68 | 0.08 | 1.08 | 0.46 | 2.30 | 1.06 | 0.08 | 1.60 | 0.46 | 3.20 |
| 30 | 0.77 | 0.08 | 1.26 | 0.46 | 2.57 | 1.15 | 0.08 | 1.71 | 0.46 | 3.40 |
| 32 | 0.88 | 0.08 | 1.39 | 0.54 | 2.89 | 1.25 | 0.08 | 1.87 | 0.54 | 3.74 |
| 34 | 1.02 | 0.08 | 1.59 | 0.54 | 3.23 | 1.35 | 0.08 | 2.02 | 0.54 | 3.99 |
| 36 | 1.16 | 0.08 | 1.78 | 0.54 | 3.56 | 1.52 | 0.08 | 2.23 | 0.54 | 4.37 |
| 40 | 1.60 | 0.13 | 2.34 | 0.63 | 4.07 | 1.92 | 0.13 | 2.73 | 0.63 | 5.41 |
| 42 | 1.85 | 0.13 | 2.68 | 0.63 | 5.29 | 2.17 | 0.13 | 3.05 | 0.63 | 5.98 |

The related construction labor and equipment spreads that follow should be capable of installing the following pipeline to platform pipe risers within the time frames listed above.

Above item numbers represent:

1) Capping off of pipeline and laying on bottom.
2) Measuring on bottom with divers to obtain exact cut-off point for riser connection.
3) Raise pipeline to lay barge, cut off and connect prefabricated riser.
4) Coat and wrap riser where required, lower back to bottom and fasten riser to jacket and platform.

T) Total of the above operations time.

Above time frames do not include fabrication of risers.

If riser is to be installed by platform crew, see related tables under section 7, "Offshore Installation of Equipment and Appurtenances."

# AVERAGE PIPELINE TO PLATFORM RISER INSTALLATION (Water Depth from 150 to 200 Feet)

CREW AND EQUIPMENT DAYS PER RISER

| O.D. Pipe Size Inches | PIPE WALL THICKNESS | | | | | | | | | |
|---|---|---|---|---|---|---|---|---|---|---|
| | to 0.500 | | | | | 0.500 to 1.000 | | | | |
| | Item Numbers | | | | | Item Numbers | | | | |
| | 1 | 2 | 3 | 4 | T | 1 | 2 | 3 | 4 | T |
| 4.500 | 0.17 | 0.08 | 0.19 | 0.25 | 0.69 | 0.27 | 0.08 | 0.30 | 0.25 | 0.90 |
| 5.500 | 0.19 | 0.08 | 0.21 | 0.25 | 0.73 | 0.29 | 0.08 | 0.33 | 0.25 | 0.95 |
| 5.563 | 0.20 | 0.08 | 0.22 | 0.25 | 0.75 | 0.31 | 0.08 | 0.35 | 0.25 | 0.99 |
| 6.625 | 0.21 | 0.08 | 0.24 | 0.25 | 0.78 | 0.38 | 0.08 | 0.42 | 0.25 | 1.13 |
| 8.625 | 0.23 | 0.08 | 0.28 | 0.29 | 0.88 | 0.46 | 0.08 | 0.53 | 0.29 | 1.36 |
| 10.750 | 0.27 | 0.08 | 0.34 | 0.29 | 0.98 | 0.58 | 0.08 | 0.68 | 0.29 | 1.63 |
| 12.750 | 0.31 | 0.08 | 0.40 | 0.29 | 1.08 | 0.63 | 0.08 | 0.75 | 0.29 | 1.75 |
| 14.000 | 0.38 | 0.08 | 0.48 | 0.38 | 1.32 | 0.80 | 0.08 | 0.95 | 0.38 | 2.21 |
| 16.000 | 0.40 | 0.08 | 0.53 | 0.38 | 1.39 | 0.83 | 0.08 | 1.01 | 0.38 | 2.30 |
| 18.000 | 0.45 | 0.08 | 0.61 | 0.38 | 1.52 | 0.88 | 0.08 | 1.10 | 0.38 | 2.44 |
| 20.000 | 0.48 | 0.13 | 0.68 | 0.42 | 1.71 | 0.90 | 0.13 | 1.15 | 0.42 | 2.60 |

The related construction labor and equipment spreads that follow should be capable of installing the following pipeline to platform pipe risers within the time frames listed above.

Above item numbers represent:

1) Capping off of pipeline and laying on bottom.
2) Measuring on bottom with divers to obtain exact cut-off point for riser connection.
3) Raise pipeline to lay barge, cut off and connect prefabricated riser.
4) Coat and wrap riser where required, lower back to bottom and fasten riser to jacket and platform.
T) Total of the above operations time.

Above time frames do not include fabrication of risers.

If riser is to be installed by platform crew, see related tables under section 7, "Offshore Installation of Equipment and Appurtenances."

# AVERAGE PIPELINE TO PLATFORM RISER INSTALLATION (Water Depth from 150 to 200 Feet)

CREW AND EQUIPMENT DAYS PER RISER

| O.D. Pipe Size Inches | PIPE WALL THICKNESS | | | | | | | | | |
|---|---|---|---|---|---|---|---|---|---|---|
| | to 0.500 | | | | | 0.500 to 1.000 | | | | |
| | Item Numbers | | | | | Item Numbers | | | | |
| | 1 | 2 | 3 | 4 | T | 1 | 2 | 3 | 4 | T |
| 22 | 0.51 | 0.13 | 0.73 | 0.42 | 1.79 | 0.93 | 0.13 | 1.23 | 0.42 | 2.71 |
| 24 | 0.56 | 0.13 | 0.82 | 0.42 | 1.93 | 0.95 | 0.13 | 1.32 | 0.42 | 2.82 |
| 26 | 0.65 | 0.13 | 1.06 | 0.50 | 2.34 | 1.02 | 0.13 | 1.51 | 0.50 | 3.16 |
| 28 | 0.72 | 0.13 | 1.12 | 0.50 | 2.47 | 1.10 | 0.13 | 1.64 | 0.50 | 3.37 |
| 30 | 0.81 | 0.13 | 1.30 | 0.50 | 2.74 | 1.19 | 0.13 | 1.75 | 0.50 | 3.57 |
| 32 | 0.92 | 0.13 | 1.43 | 0.58 | 3.06 | 1.29 | 0.13 | 1.91 | 0.58 | 3.91 |
| 34 | 1.06 | 0.13 | 1.63 | 0.58 | 3.40 | 1.40 | 0.13 | 2.16 | 0.58 | 4.27 |
| 36 | 1.20 | 0.13 | 1.82 | 0.58 | 3.73 | 1.56 | 0.13 | 2.27 | 0.58 | 4.54 |
| 40 | 1.65 | 0.17 | 2.38 | 0.67 | 4.87 | 1.96 | 0.17 | 2.77 | 0.67 | 5.58 |
| 42 | 1.89 | 0.17 | 2.72 | 0.67 | 5.45 | 2.21 | 0.17 | 3.09 | 0.67 | 6.14 |

The related construction labor and equipment spreads that follow should be capable of installing the following pipeline to platform pipe risers within the time frames listed above.

Above item numbers represent:

1) Capping off of pipeline and laying on bottom.
2) Measuring on bottom with divers to obtain exact cut-off point for riser connection.
3) Raise pipeline to lay barge, cut off and connect prefabricated riser.
4) Coat and wrap riser where required, lower back to bottom and fasten riser to jacket and platform.

T) Total of the above operations time.

Above time frames do not include fabrication of risers.

If riser is to be installed by platform crew, see related tables under section 7, "Offshore Installation of Equipment and Appurtenances."

# AVERAGE PIPELINE TO PLATFORM RISER INSTALLATION (Water Depth from 200 to 250 Feet)

CREW AND EQUIPMENT DAYS PER RISER

| O.D. Pipe Size Inches | PIPE WALL THICKNESS | | | | | | | | | |
|---|---|---|---|---|---|---|---|---|---|---|
| | to 0.500 | | | | | 0.500 to 1.000 | | | | |
| | Item Numbers | | | | | Item Numbers | | | | |
| | 1 | 2 | 3 | 4 | T | 1 | 2 | 3 | 4 | T |
| 4.500 | 0.21 | 0.13 | 0.24 | 0.29 | 0.87 | 0.31 | 0.13 | 0.35 | 0.29 | 1.08 |
| 5.500 | 0.23 | 0.13 | 0.26 | 0.29 | 0.91 | 0.33 | 0.13 | 0.38 | 0.29 | 1.13 |
| 5.563 | 0.24 | 0.13 | 0.27 | 0.29 | 0.93 | 0.35 | 0.13 | 0.40 | 0.29 | 1.17 |
| 6.625 | 0.25 | 0.13 | 0.29 | 0.29 | 0.96 | 0.42 | 0.13 | 0.47 | 0.29 | 1.31 |
| 8.625 | 0.27 | 0.13 | 0.33 | 0.33 | 1.06 | 0.50 | 0.13 | 0.58 | 0.33 | 1.54 |
| 10.750 | 0.31 | 0.13 | 0.39 | 0.33 | 1.16 | 0.54 | 0.13 | 0.73 | 0.33 | 1.73 |
| 12.750 | 0.35 | 0.13 | 0.45 | 0.33 | 1.26 | 0.67 | 0.13 | 0.80 | 0.33 | 1.93 |
| 14.000 | 0.42 | 0.13 | 0.52 | 0.42 | 1.49 | 0.84 | 0.13 | 0.99 | 0.42 | 2.38 |
| 16.000 | 0.44 | 0.13 | 0.57 | 0.42 | 1.56 | 0.88 | 0.13 | 1.05 | 0.42 | 2.48 |
| 18.000 | 0.49 | 0.13 | 0.65 | 0.42 | 1.69 | 0.93 | 0.13 | 1.14 | 0.42 | 2.62 |
| 20.000 | 0.52 | 0.17 | 0.69 | 0.46 | 1.84 | 0.95 | 0.17 | 1.19 | 0.46 | 2.77 |

The related construction labor and equipment spreads that follow should be capable of installing the following pipeline to platform pipe risers within the time frames listed above.

Above item numbers represent:

1) Capping off of pipeline and laying on bottom.
2) Measuring on bottom with divers to obtain exact cut-off point for riser connection.
3) Raise pipeline to lay barge, cut off and connect prefabricated riser.
4) Coat and wrap riser where required, lower back to bottom and fasten riser to jacket and platform.

T) Total of the above operations time.

Above time frames do not include fabrication of risers.

If riser is to be installed by platform crew, see related tables under section 7, "Offshore Installation of Equipment and Appurtenances."

# AVERAGE PIPELINE TO PLATFORM RISER INSTALLATION (Water Depth from 200 to 250 Feet)

CREW AND EQUIPMENT DAYS PER RISER

| O.D. Pipe Size Inches | PIPE WALL THICKNESS | | | | | | | | | |
|---|---|---|---|---|---|---|---|---|---|---|
| | to 0.500 | | | | | 0.500 to 1.000 | | | | |
| | Item Numbers | | | | | Item Numbers | | | | |
| | 1 | 2 | 3 | 4 | T | 1 | 2 | 3 | 4 | T |
| 22 | 0.55 | 0.17 | 0.77 | 0.46 | 1.95 | 0.97 | 0.17 | 1.27 | 0.46 | 2.87 |
| 24 | 0.60 | 0.17 | 0.86 | 0.46 | 2.09 | 0.99 | 0.17 | 1.36 | 0.46 | 2.98 |
| 26 | 0.69 | 0.17 | 1.10 | 0.54 | 2.50 | 1.06 | 0.17 | 1.55 | 0.54 | 3.32 |
| 28 | 0.76 | 0.17 | 1.16 | 0.54 | 2.63 | 1.15 | 0.17 | 1.68 | 0.54 | 3.54 |
| 30 | 0.85 | 0.17 | 1.34 | 0.54 | 2.90 | 1.23 | 0.17 | 1.79 | 0.54 | 3.73 |
| 32 | 0.96 | 0.17 | 1.47 | 0.63 | 3.23 | 1.33 | 0.17 | 1.95 | 0.63 | 4.08 |
| 34 | 1.10 | 0.17 | 1.67 | 0.63 | 3.57 | 1.44 | 0.17 | 2.10 | 0.63 | 4.34 |
| 36 | 1.24 | 0.17 | 1.86 | 0.63 | 3.90 | 1.60 | 0.17 | 2.31 | 0.63 | 4.71 |
| 40 | 1.69 | 0.21 | 2.42 | 0.71 | 5.03 | 2.00 | 0.21 | 2.81 | 0.71 | 5.73 |
| 42 | 1.93 | 0.21 | 2.76 | 0.71 | 5.61 | 2.25 | 0.21 | 3.13 | 0.71 | 6.30 |

The related construction labor and equipment spreads that follow should be capable of installing the following pipeline to platform pipe risers within the time frames listed above.

Above item numbers represent:

1) Capping off of pipeline and laying on bottom.
2) Measuring on bottom with divers to obtain exact cut-off point for riser connection.
3) Raise pipeline to lay barge, cut off and connect prefabricated riser.
4) Coat and wrap riser where required, lower back to bottom and fasten riser to jacket and platform.

T) Total of the above operations time.

Above time frames do not include fabrication of risers.

If riser is to be installed by platform crew, see related tables under section 7, "Offshore Installation of Equipment and Appurtenances."

# PRODUCTIVITY AND PERCENTAGE EFFICIENCY FACTORS

To correctly apply the time frames in the preceding tables, consideration must be given to the effects of labor productivity and of wind, wave and current action.

For estimating the time involved in the installation of submerged pipelines, the following labor, wind, wave and current efficiency percentages should be applied to the preceding time frame tables according to the conditions encountered on a particular project. See pages 150-151 for an example of the application of these factors to the time frame tables.

Productivity factors to be applied to number of single random joints of pipelay per day for labor are as follows:

LABOR PRODUCTIVITY FACTORS

| Type of Productivity | Percentage Range |
|---|---|
| Excellent | 0.901 through 1.000 |
| Very Good | 0.801 through 0.900 |
| Average | 0.601 through 0.800 |
| Low | 0.401 through 0.600 |
| Very Low | 0.001 through 0.400 |

The following production elements should be considered in the application of a labor productivity percentage:

1. General Economy
2. Project Supervision
3. Labor Conditions
4. Job Conditions
5. Equipment
6. Weather

For full description of above elements and example of obtaining labor productivity percentage, see pages xi-xiv.

Productivity factors for laying concrete weight coated pipe under certain average wind conditions are as follows:

WIND PRODUCTIVITY FACTORS

| Description | Wind (mph) | Percent Efficiency |
|---|---|---|
| Calm | 0 - 1 | 100 |
| Light Air | 1 - 3 | 100 |
| Slight Breeze | 4 - 7 | 95 |
| Gentle Breeze | 8 - 12 | 90 |
| Moderate Breeze | 13 - 18 | 75 |
| Fresh Breeze | 19 - 24 | 50 |
| Strong Breeze | 25 - 31 | 30 |

The above wind descriptions and speeds are taken from Beaufort's Wind Scale, as modified for steam.

Efficiency percentages may fluctuate slightly, depending on type of sea-going equipment used, presence of wind breakers when necessary, experience of personnel, etc.

Above efficiency percentages are based on actual time and motion studies made from pipelay operations under average wind speeds as listed above.

Productivity factors for laying concrete weight coated pipe under certain average current conditions are as follows:

CURRENT PRODUCTIVITY FACTORS

| Average Total Current (feet per second) | Percent Efficiency |
|---|---|
| 0.0 to 0.5 | 100 |
| 0.5 to 1.0 | 97 |
| 1.0 to 2.0 | 95 |
| 2.0 to 2.5 | 90 |
| 2.5 to 3.0 | 85 |
| 3.0 to 3.5 | 78 |
| 3.5 to 4.0 | 70 |
| 4.0 to 5.0 | 65 |

These currents are of sufficient magnitude to influence underwater pipelay operations.

Average total current speeds represent total average currents of tidal current, density current and wind drift current.

Above efficiency percentages are based on actual time and motion studies made from pipelay operations under the average current speeds as listed.

Productivity factors for various pieces of sea-going equipment under certain wave conditions are as follows:

WAVE PRODUCTIVITY FACTORS

| Equipment and Type of Operations | WAVE HEIGHT IN FEET AND PERCENTAGE EFFICIENCY FOR: | | | | | |
|---|---|---|---|---|---|---|
| | Safe Efficient Operations | | Marginal Operations | | Dangerous and/or Inefficient Operations | |
| | Wave Height | Percent Efficiency | Wave Height | Percent Efficiency | Wave Height | Percent Efficiency |
| Deep Sea Tug: | — | — | — | — | — | — |
| Towing lay barge | 0 - 4 | 100 - 70 | 4 - 6 | 70 - 50 | 6 or greater | 50 - 20 |
| Towing material barge | 0 - 4 | 100 - 70 | 4 - 6 | 70 - 50 | 6 or greater | 50 - 20 |
| Working lay barge | 0 - 2 | 100 - 70 | 2 - 3 | 70 - 40 | 3 or greater | 40 - 10 |
| Working material barge | 0 - 2 | 100 - 70 | 2 - 3 | 70 - 40 | 3 or greater | 40 - 10 |
| Crew Boats (60 to 90 feet long): | — | — | — | — | — | — |
| Underway | 0 - 8 | 100 - 80 | 8 - 15 | 80 - 40 | 15 or greater | 40 - 10 |
| Loading or unloading crews | 0 - 3 | 100 - 70 | 3 - 5 | 70 - 50 | 5 or greater | 50 - 20 |
| Lay or Derrick Barge: | — | — | — | — | — | — |
| Small barge—underway | 0 - 2 | 100 - 70 | 2 - 3 | 70 - 50 | 3 or greater | 50 - 20 |
| Large barge—underway | 0 - 3 | 100 - 70 | 3 - 5 | 70 - 50 | 5 or greater | 50 - 20 |
| Small barge—working | 0 - 2 | 100 - 70 | 2 - 3 | 70 - 40 | 3 or greater | 40 - 10 |
| Large barge—working | 0 - 3 | 100 - 70 | 3 - 4 | 70 - 40 | 4 or greater | 40 - 10 |

Wave heights used are the average maximum for the work intended.

"Wave height" is defined as the average height of the highest one-third of the wave passing a stationary location during a given observation period.

Wave height limits given above are not rigid but will vary to some extent with locality, local wind conditions, experience of personnel, etc.

The above wave limits will fluctuate slightly, depending on the position of the vessel relative to the direction of wave approach; it is recommended that the vessel be oriented with its bow into prevailing seas while working.

# LAY BARGE LABOR AND EQUIPMENT SPREADS

## General Notes

In keeping with the average pipelay tables (pages 120-140), the following items must be considered when applying the lay barge labor and equipment spreads:

1) These labor and equipment spreads are based on welds made by the manual shielded metal-arc method.
2) Welding machines should be of the size and type suitable for the work.
3) Welding machines are to be operated within the amperage and voltage ranges recommended for each size and type of electrode.
4) Remove all foreign matter from the beveled ends before welding. Clean with hand or power tools.
5) Pipe ends must have proper bevels for the work intended.
6) The alignment of abutting pipe ends should be such as to minimize the offset between pipe surface. Hammering of pipe to obtain proper line-up should be held to a minimum.
7) Internal line-up clamps should be used where possible. Use external line-up clamps when it is impractical or impossible to use an internal line-up clamp.
8) Brush and clean the finished weld.
9) Only qualified welders should be used.
10) Where and if required, apply heat treatment—preheating and postheating.
11) Apply radiographic inspection of welds where required.
12) Clean and remove all dirt, mill scale, rust, welding scale and all other foreign material prior to coating, wrapping and concrete weight coating of pipe joints.
13) Handle concrete weight coated pipe with equipment designed to prevent damage to the coating at all times.
14) Check all joint coating and wrapping for holidays with a high-voltage electric holiday detector.

# OCEAN AND SEA SUBMERGED PIPE INSTALLATION
## Aligning, Welding and Installing Operations

**Labor Crew**

| Personnel Description | Number of Men |
|---|---|
| Superintendent or Captain | 1 |
| Assistant Superintendent | 1 |
| Barge Foreman | 2 |
| Welder Foreman | 2 |
| Anchor Foreman | 2 |
| Pot/Mixer Foreman | 2 |
| Dope Foreman | 2 |
| Welder | 18 |
| Welder Helper | 18 |
| Rack Operator | 2 |
| Spacer | 4 |
| Clampman | 2 |
| Buffer | 4 |
| Crane Operator—To 600 Tons | 2 |
| Crane Operator—100 Tons | 2 |
| Hoist Operator | 4 |
| Anchor Hoist Operator | 4 |
| Tension Device Operator | 2 |
| Pot Fireman | 2 |
| Mixer Operator | 2 |
| Laborer | 24 |
| Field Engineer | 1 |
| X-Ray Technician | 2 |
| X-Ray Helper | 4 |
| Clerk | 2 |
| Radio Operator/Technician | 1 |
| Assistant Clerk/Radio Operator | 1 |
| Total Crew | 113 |

The above labor crew is for the on-site supervision and installation of submerged pipelines and should be ample for the work intended as outlined in the average pipelay tables, (pages 120-140).

Other labor crews in this section must be added for a complete working spread.

# OCEAN AND SEA SUBMERGED PIPE INSTALLATION

### Lay Barge Equipment Spread

| Equipment Description | Number of Units |
|---|---|
| Pipelay Barge with Incline Pipe Ramp | 1 |
| Crawler Crane—150-Ton | 1 |
| Crawler Crane—100-Ton | 1 |
| Anchor Handling Winches | 8 |
| Anchors—12,000 Pounds Each± | 8 |
| 1½" Anchor Wire Rope—300 L.F. Each | 8 |
| Pipe Rack | 1 |
| Pipe Stalking Device—Remote Control | 1 |
| Pipe Conveyor | 1 |
| Pipe Handling Device | 5 |
| Stinger—Prefabricated or Articulate | 1 |
| Tension Device | 2 |
| Welding Stations | 4 |
| Welding Machines—400 Amp | 16 |
| Field Joint Station | 1 |
| Dope Pot—10 Barrel | 2 |
| Concrete Mixer—Small | 2 |
| X-Ray Equipment (Complete Set-up) | 1 |
| Air Compressor—600 CFM | 2 |
| Holiday Detector | 3 |
| Water Pump—4" | 2 |
| Water Pump—6" | 4 |

In keeping with the average pipelay tables the lay barge should be equipped with the items above.

If combination lay/derrick barge with large revolving crane is to be utilized, delete the 150-ton crane from the above.

If pipeline is to be buried using the jetting process, add jetting equipment to the above.

The following jetting equipment should be capable of burying pipe up to 12 feet deep:

| Equipment Description | Number of Units |
|---|---|
| Jetting Pumps—2000 GPM @ 1000 PSI | 4 |
| Dredge Pump—20-Inch | 1 |
| Suction and Jetting Ladder | 1 |
| Jetting Sled | 1 |

# OCEAN AND SEA SUBMERGED PIPE INSTALLATION
## Lay or Derrick Barge Maintenance Operation

**Labor Crew**

| Personnel Description | Number of Men |
|---|---|
| Chief Engineer | 1 |
| Shift Engineer | 2 |
| Machinist | 2 |
| Electrician | 2 |
| Engineer | 2 |
| Oiler | 2 |
| First Aid Man | 1 |
| Total Crew | 12 |

The above labor crew is for maintaining the lay barge and assuring that all its mechanical and electrical parts function properly and that emergency first aid is available.

## Lay, Derrick or Quarter Barge Quartering and Catering

**Labor Crew**

| Personnel Description | Number of Men |
|---|---|
| Chief Cook | 2 |
| Second Cook | 2 |
| Chief Stewart | 1 |
| Mess Boy | 2 |
| Janitor | 4 |
| Laundry Man | 2 |
| Total Crew | 13 |

Duties of the above labor crew consist of preparing and serving meals, house cleaning, and maintaining laundry facilities for the lay barge personnel.

These duties can be performed on either the lay barge or a special quartering barge, depending on the available facilities.

# OCEAN AND SEA SUBMERGED PIPE INSTALLATION
## Crew Boat Spread

### Labor Crew

| Personnel Description | Number of Men |
|---|---|
| Captain | 1 |
| Mate/Engineer | 1 |
| Seaman | 1 |
| Total Crew | 3 |

### Equipment Spread

| Equipment Description | Number of Units |
|---|---|
| Crew Boat—65′ ± | 1 |

The above labor crew and equipment spread are for the purpose of hauling personnel between shore and installation site.

When and where feasible, the spread can also be utilized for handling and hauling miscellaneous light supplies such as welding rods and food items.

The size of the total crews and the location of the project govern the size and number of crew boats required.

# OCEAN AND SEA SUBMERGED PIPE INSTALLATION
## Diver Labor and Equipment Spread

### Labor Crew

| Personnel Description | Number of Men |
|---|---|
| Diver Supervisor | 1 |
| Divers | 4 |
| Diver Tenders | 4 |
| Total Crew | 9 |

### Equipment Spread

| Equipment Description | Number of Units |
|---|---|
| Decompression Chamber | 2 |
| High Compression Compressor | 2 |
| Diving Gear | 4 |
| Oxygen Quads | 4 |
| Lot of Small Hand Tools | 1 |

The above labor crew and equipment spread are ample for all underwater tasks such as bottom, pipeline, weld and obstruction inspections.

Adjustments to the above labor crew and equipment spread may be required, depending on water depth and quantity and type of operations to be performed.

# OCEAN AND SEA SUBMERGED PIPE INSTALLATION
## Work or Supply Boat Spread

### Operating Labor Crew

| Personnel Description | Number of Men |
|---|---|
| Captain | 1 |
| First Mate | 1 |
| Chief Engineer | 1 |
| First Engineer | 1 |
| Electrician | 1 |
| Oiler | 1 |
| Seaman | 4 |
| Cook | 1 |
| Total Crew | 11 |

### Equipment Spread

| Equipment Description | Number of Units |
|---|---|
| Work Boat | 1 |

The above labor crew and equipment spread are for maintaining and operating a work or supply boat necessary for hauling supplies or assisting in certain special operations such as support for survey crews, etc.

The size, horsepower, and rating of the boat are governed by the duties it is to perform and the waters it is to be performing in.

For small limited-capacity boats working fairly close to shore in relatively calm waters, the above labor crew can be adjusted downward.

# OCEAN AND SEA SUBMERGED PIPE INSTALLATION
## Survey Spread

### Labor Crew

| Personnel Description | Number of Men |
|---|---|
| Chief Survey Engineer | 1 |
| Survey Engineers | 3 |
| Helpers | 6 |
| Total Crew | 10 |

### Equipment Spread

| Equipment Description | Number of Units |
|---|---|
| Work Boat (Fully Manned) | 1 |
| Survey Equipment | 1 |

The above labor crew and equipment spread are for surveying and laying out proposed pipelines.

The time required for this operation is governed by the length and number of required lines.

See the work or supply boat spread on page 158 for fully manned boat requirements.

# OCEAN AND SEA SUBMERGED PIPE INSTALLATION
## Tug Spread

### Labor Crew

| Personnel Description | Number of Men |
|---|---|
| Captain | 1 |
| First Mate | 1 |
| Chief Engineer | 1 |
| First Engineer | 1 |
| Seaman | 4 |
| Oiler | 1 |
| Cook | 1 |
| Total Crew | 10 |

### Equipment Spread

| Equipment Description | Number of Units |
|---|---|
| Tug | 1 |

The above labor crew and equipment spread are for maintaining and operating a tug of sufficient size to handle and tow the lay barge and material barges and to handle, where and when required, the anchors of these barges.

The size, horsepower and rating of the tug are governed by the equipment it is to handle and the waters it is to be towing or working in.

For small, limited-capacity tugs working relatively close to shore in relatively clam waters, the above labor crew can be adjusted downward.

# OCEAN AND SEA SUBMERGED PIPE CLEANING AND TESTING

**Labor Crew**

| Personnel Description | Number of Men |
|---|---|
| Pipefitter Welder Foreman | 2 |
| Operator | 2 |
| Oiler | 2 |
| Welder | 8 |
| Welder Helper | 8 |
| Pipefitter | 8 |
| Pipefitter Helper | 8 |
| Laborer | 12 |
| Total Crew | 50 |

The above labor crew is ample for pig cleaning and hydrostatic testing of pipelines.

If partial pigging and testing of pipelines are performed during pipelay operations, a portion of the lay barge crew can be used for this operation and the above labor crew will not be required.

If pigging and testing are to be performed after complete installation of the pipelines, the lay barge crew for pipelay operations can be reduced to the above labor crew.

The time required to pig and test line is determined by length of time required to set up pig launcher and receiver and length and size of pipeline to be tested; this must be estimated for the individual project.

# OCEAN AND SEA SUBMERGED PIPE CLEANING AND TESTING

### Equipment Spread

| Equipment Description | Unit | Number of Units |
|---|---|---|
| Pig Launcher | Each | 1 |
| Pig Trap | Each | 1 |
| Pigs | Each | 4 |
| Misc. Pipe, Valves and Fittings | Lot | 1 |
| Anchor Forgings and Tie Downs | Lot | 1 |
| Miscellaneous Gauges | Lot | 1 |
| Radios (Walkie Talkies) | Each | 2 |

The above equipment spread should be added to the lay barge equipment spread if pigging and testing operations are to be conducted as pipelines are being installed.

If pigging and testing of pipelines are to be conducted after pipelines are completely installed, and if the availability and transportation cost of obtaining the above equipment is less than the daily cost of this added equipment (which would have been assessed against the lay barge equipment spread had it been included for the lay barge's total tour of duty), then the above equipment should be added to the lay barge for these operations for the length of time required.

The time requirements for the above equipment to pig and test line are governed by length of time required to set up pig launcher and receiver and length and size of pipeline to be tested; this must be estimated for the individual project.

# OCEAN AND SEA SUBMERGED PIPE INSTALLATION
## Sandblast and Paint Labor and Equipment Spread

**Labor Crew**

| Personnel Description | Number of Men |
|---|---|
| Paint Foreman | 1 |
| Blasters and Painters | 2 |
| Pot Man/Helper | 3 |
| Operator | 1 |
| Total Crew | 7 |

**Equipment Spread**

| Equipment Description | Number of Units |
|---|---|
| Air Compressor—600 CFM | 1 |
| Sand Pot and Spray Equipment | 1 |
| Air Hose—100 Feet | 1 |
| Sand Blast Hose—200 Feet | 1 |
| Lot of Small Tools | 1 |

The above labor crew and equipment spread are for cleaning (by blasting) and painting of pipe risers above the splash zone and should be added to the lay barge spread only for the time required for this operation. If possible, the above crew should be composed of other craftsmen already aboard the lay barge.

# Section Five

# DREDGING, BLASTING AND JETTING

Before an estimate is made on dredged, blasted or jetted excavation, it is well to know the kind of soil that may be encountered. Soils vary greatly in character, and no two are exactly alike. For this reason, we have classified soil into the following four groups according to the difficulty experienced in excavating it:

*Light Soil*—Soil, such as sand, which is easily removed with drag bucket and requires no loosening.

*Medium Soil*—Soil easily removed but which requires preliminary loosening with bucket prior to removal. This type of earth is usually classified as ordinary soil and loam.

*Heavy Soil*—This type of soil can be loosened with the drag bucket, but this is sometimes difficult. Hard and compact loam containing gravel, small stones and boulders (stiff clay or compact gravel are good examples of this type).

*Rock*—Requires blasting before removal. May be divided into different grades such as hard, soft or medium.

Backfilling is simply the replacement of excavations with pre-excavated materials or select materials that have been hauled in.

The time frames in the tables in this section are based on the following conditions:

1) The time frames listed are for quantities of compact measurement.
2) The equipment is working in the optimum depth of cut for maximum efficiency.
3) There are no delays—the equipment is working a full 60 minutes each hour.
4) The dragline is making a 90° swing before unloading.
5) The proper type bucket is being used for the job.
6) The dragline is being used within the working radius recommended by the manufacturer for machine stability.

We have attempted to simplify requirements and data pertaining to the use of explosives in trenching. We cannot put together a complete requirement and time frame data system for all conditions that might arise, since there are too many unknowns involved. We feel that the information in this section will be useful for estimating the majority of projects one might encounter.

Onshore magazine storage facilities for storage of explosives are not included and must be added to the estimate for blasting.

The cost of all local explosive fees, licenses and permits also must be added.

# DREDGING—EXCAVATING AND BACKFILL

## Spread Hours Per 100 Cubic Yards

| Dragline Bucket Capacity | SPREAD HOURS REQUIRED FOR EXCAVATING SOIL CONDITIONS | | | |
|---|---|---|---|---|
| | Light | Medium | Heavy | Backfill |
| 2 Cubic Yards ...... | 0.70 | 1.30 | 1.70 | 1.60 |
| 6½ Cubic Yards ...... | 0.30 | 0.56 | 0.73 | 1.20 |
| 7½ Cubic Yards ........ | 0.18 | 0.34 | 0.45 | 1.00 |
| 10 Cubic Yards ........ | 0.17 | 0.32 | 0.42 | 0.65 |

Excavating spread hours include operation of excavating equipment and dumping of excavated materials on side or loading on material barge for hauling and dumping at sea.

Spread hours do not include hauling and dumping of excavated materials at sea or hauling of select material from outside source for backfill. Add for these operations if required or necessary.

If excavated or backfill quantity is less than 1000 cubic yards add 30 percent to above spread hours.

## Jetting Hours Required Per 100 Linear Feet

| Water Depth | SPREAD HOURS REQUIRED FOR | | |
|---|---|---|---|
| | Light Soil | Medium Soil | Heavy Soil |
| To 19 Feet .................. | 1.00 | 1.25 | 1.75 |
| 19 To 50 Feet .................. | 1.50 | 1.75 | 2.25 |
| 50 To 100 Feet .................. | 2.00 | 2.50 | 3.25 |
| 100 To 200 Feet .................. | 3.00 | 3.50 | 4.00 |

See jetting operation spread (page 172) to be added to dredging spread in application of above time frames.

# DREDGING—EXCAVATING AND BACKFILL

### Labor Crew

| Personnel Description | Number of Men Required |
|---|---|
| Barge Foreman | 2 |
| Anchor Operator | 2 |
| Crane Operator | 2 |
| Oiler | 2 |
| Swamper | 4 |
| Deck Hands/Laborers | 12 |
| Total Crew | 24 |

### Equipment Spread

| Equipment Description | Unit | Number of Units Required |
|---|---|---|
| Work Barge | Each | 1 |
| Crane With Drag Line and Bucket | Each | 1 |
| Double Drum Anchor Winches | Each | 4 |
| 1½" Wire Rope Anchor Cable | L.F. | 12,000 |
| Anchors—10,000 Pounds | Each | 8 |

Above labor crew is based on working a two-shift twenty-four hour day.

Labor crew and equipment spread are ample for dredging and backfilling offshore materials in the quantities listed in the preceding time table and according to procedures outlined in the section introduction, page 165.

# AVERAGE TIME REQUIRED FOR TRENCHING BY BLASTING

**Crew Days Required Per 1,000 Linear Feet**

| Operation Description | CREW DAYS REQUIRED FOR | | |
|---|---|---|---|
| | Surf Zone Sandy Beach | Reef or Rock | Mud Flats |
| Move on Location | 2 | 2 | 1 |
| Receive and Store Materials | 1 | 2 | 1 |
| Prepare and Load Charges | 5 | 6 | 3 |
| String Charges | 2 | 3 | 1 |
| Tie-in Primacord and Blast | 1 | 2 | 1 |
| Total Crew Days | 11 | 15 | 7 |

The related construction labor crew, equipment spread and required materials tables that follow in this section should be capable of opening 1,000 linear feet of ditch by blasting, working a 12-hour single-shift day, for the number of crew days listed above.

The above items include the following:

*Move on Location*—Move on location and prepare magazine to receive materials.

*Receive and Store Materials*—Receiving, segregating, and storing of blasting and miscellaneous materials in magazine.

*Prepare and Load Charges*—Preparing by mixing ingredients and loading of charges.

*String Charges*—Placing and anchoring of charges in correct position for blasting.

*Tie-in Primacord and Blast*—Tie primacord to charges, hook-up and blast.

Above time frames are based on working in protected waters with very little interference from wind, waves and currents. If conditions other than these exist, see Productivity and Percentage Efficiency Factors, pages 147-149.

# TRENCH BLASTING

**Single 12-Hour Shift Labor Crew**

| Personnel Description | NUMBER OF MEN REQUIRED FOR | | |
|---|---|---|---|
| | Surf Zone Sandy Beach | Reef or Rock | Mud Flats |
| Superintendent | 1 | 1 | 1 |
| Barge Foreman | 0 | 1 | 1 |
| Blasting Foreman | 1 | 1 | 1 |
| Powderman/Laborer | 4 | 8 | 6 |
| Welder | 1 | 1 | 1 |
| Burner | 1 | 1 | 1 |
| Operator | 2 | 2 | 1 |
| Oiler | 2 | 2 | 1 |
| Divers | 1 | 1 | 1 |
| Truck Driver | 1 | 1 | 1 |
| Total | 14 | 19 | 15 |

Quarter and catering personnel must be added if required; see labor crew as listed on page 155.

If fully manned derrick or lay barge is available at time of blasting operations, the following personnel can be eliminated from the above:

i) Barge Foreman
ii) Welder
iii) Burner
iv) Operator
v) Oiler
vi) Divers

See various labor crews as listed under "Ocean and Sea Pipelines," Section Four, for required make-up of above labor.

# TRENCH BLASTING

## Equipment Spread

| Equipment Description | NUMBER OF UNITS FOR | | | |
|---|---|---|---|---|
| | Unit | Surf Zone Sandy Beach | Reef or Rock | Mud Flats |
| Work Barge With Anchors | Each | 0 | 1 | 1 |
| Tug Boat (Fully Manned) | Each | 0 | 1 | 1 |
| Light Crawler Crane | Each | 0 | 1 | 1 |
| Small Crew Boat (Manned) | Each | 0 | 1 | 1 |
| Generator | Each | 2 | 2 | 2 |
| Welding Machine | Each | 2 | 2 | 2 |
| Diver Equipment | Lot | 0 | 1 | 1 |
| Water Pump | Each | 2 | 2 | 2 |
| Bulldozer | Each | 1 | 1 | 0 |
| Side Boom | Each | 1 | 0 | 0 |
| Flat Bed Truck | Each | 1 | 1 | 1 |

Small tools and consumable supplies as required must be added to the above spread.

Quartering and catering facilities must be added if required.

If fully equipped derrick or lay barge is available at time of blasting operations, the following equipment can be eliminated from the above:

i) Work Barge
ii) Tug Boat
iii) Light Crawler Crane
iv) Small Crew boat
v) Generator
vi) Welding Machine
vii) Diver Equipment
viii) Water Pump

See various equipment spreads listed under "Ocean and Sea Pipelines," Section Four, for required make-up of above equipment.

# TRENCH BLASTING

**Material Required Per 1,000 Linear Feet**
**Two-Shot Operation**

| Material Description | Unit | MATERIAL REQUIRED Surf Zone Sandy Beach | Reef or Rock | Mud Flats |
|---|---|---|---|---|
| Nitromethene | lbs. | 21,500 | 36,500 | 16,500 |
| Dietheltriamine | lbs. | 1,500 | 3,000 | 1,000 |
| Amonimum Nitrate | lbs. | 21,600 | 25,500 | 16,000 |
| American Primacord—100 Grain | lf. | 7,000 | 7,000 | 7,000 |
| Blasting Caps—No. 8 | ea. | 50 | 50 | 50 |
| Polyethelene Booster Bottles | ea. | 270 | 250 | 200 |
| Bangalors—6" | ea. | 270 | 250 | 200 |
| Panduit Guns | ea. | 8 | 8 | 6 |
| Panduit Straps | ea. | 4,000 | 4,000 | 3,000 |
| ¾" Wire Cable | lf. | 2,200 | 2,200 | 1,650 |
| ¾" Cable Clamps | ea. | 810 | 810 | 600 |
| Shooting Line (#16-2 Stran Elec. Wire) | lf. | 3,000 | 3,000 | 3,000 |
| Concrete Blocks—12" × 12" × 12" | ea. | — | 150 | — |

The above materials are required for trench blasting in the type of soil conditions as listed and should be the maximum required to obtain the following:

*Surf Zone*—Sandy soil, pipe is to be pulled out from shore. Material should be ample to blast trench for pipe up to 36-inch diameter and allow for a minimum of 10 feet cover.

*Reef or Rock*—Coral or rock, in water depth to 60 feet. Material should be ample to blast trench for pipe up to 36-inch diameter and allow for a minimum of 10 feet cover.

*Mud Flats*—Mud, in water depth to 30 feet. Material should be ample to blast trench for pipe up to 36-inch diameter and allow for a minimum of 10 feet cover.

The above listed materials are not classified as explosives until mixed together.

# JETTING OPERATION

If burying of pipe by the jetting process is required, add the following labor and equipment to the dredging labor crew and equipment spread.

For burying of pipe from 2-inch to 20-inch O.D. in water depths to 19 feet, with a maximum coverage of 6 feet, add the following:

| Item Description | Quantity |
|---|---|
| Personnel | — |
| Pump Operator | 2 |
| Laborer | 8 |
| Equipment | — |
| Jetting Pump—900 GPM @ 300 PSI | 2 |
| Dredge Pump—12 Inch | 1 |
| Suction and Jetting Ladder | 1 |
| Spuds and Hoist (Set) | 1 |

For burying of pipe from 2-inch to 36-inch O.D. in water depths to 200 feet, with a maximum cover of 12 feet, add the following:

| Item Description | Quantity |
|---|---|
| Personnel | — |
| Pump Operator | 4 |
| Laborer | 12 |
| Equipment | — |
| Jetting Pump—2000 GPM @ 1000 PSI | 4 |
| Dredge Pump—20 Inch | 1 |
| Suction and Jetting Ladder | 1 |
| Jetting Sled | 1 |

See jetting time frames, page 166, for application of above spreads.

# Section Six

# OFFSHORE STRUCTURES

The purpose of this section is to outline various operations involved—and to list the time frames, labor crews and equipment spreads that may be required—in the installation of shoreline wharfs and jetties, offshore drilling and production platforms, offshore berthing and docking facilities and single-point mooring terminals.

Many items affect the cost and time required in the installation of offshore structures. Assuming that the proper skilled craftsmen and adequate equipment spreads are used to their best advantage, the major items that must be considered, evaluated and applied against the time frames which follow in this section are those of labor productivity and wind, wave and current conditions.

Based on one hundred percent productivity, we have included time frames for the various operations involved in the installation of offshore structures. By applying the included productivity factors for labor, wind, wave and current conditions against these time frames, it becomes a simple matter to arrive at a total installation time requirement. When installing offshore structures with a full crew and equipment spread, certain operations can be performed simultaneously. When preparing the estimate, an itemized time requirement list should be made, segregating the major items of work to be performed from those supplemental operations that can be accomplished concommitantly with the major items. The total labor and equipment spreads should be assessed against the project only for the time required to perform the major items of work.

The following crew tables represent a double shift working a 24-hour day, utilizing the equipment spreads full time.

Any available deck space on the derrick barge should be used for hauling and storing as much of the structures as possible; material or cargo barges, of the proper sizes and capacities, should be added to the spread as needed.

It is assumed that the derrick barge and tugs will be equipped with the proper short-wave radio communication equipment as required, and that sufficient walkie-talkie radios are available for communicating between the various personnel.

To arrive at the total direct daily cost of any labor and equipment spread:

1) Determine the daily rate, including all fringes, of each craft or position and multiply each rate by the number of men required for that position.
2) Determine the daily rate of each rigged-up piece of equipment, including all its components.
3) Determine the daily cost of fuel, oil and grease required to run the spread and add this cost to the spread.
4) Determine the cost of small tools and consumable supplies required to support the spread for the project, prorate on a daily basis, and add the prorated cost to the daily spread cost.
5) If quartering and catering of personnel are required, estimate the cost per man day, multiply by the number of men in the crew, and add daily rate for the total crew to the spread.
6) Summarize the total cost of (1) through (5) above to arrive at the total direct cost of the spread per working day.

To determine the daily direct cost of a spread for a non-working or stand-by day, simply reduce the working day spread cost by the cost of the fuel, oil and grease and the consumable supplies that would normally be used when working.

# PICK AND SET JACKETS
## Water Depths to 150 Feet

TOTAL SPREAD TIME REQUIRED IN HOURS

| Operation Description | TIME REQUIRED FOR FOLLOWING TONS | | | |
|---|---|---|---|---|
| | 0-200 | 201-300 | 301-400 | 401-500 |
| Set Derrick Barge in Position | 4.00 | 4.00 | 4.00 | 4.00 |
| Set Material Barge in Position | 1.00 | 1.00 | 1.00 | 1.00 |
| Break Loose Tie-Down | 2.50 | 2.75 | 3.00 | 3.25 |
| Rig Jacket with Slings and Spreaders | 0.50 | 0.50 | 0.65 | 0.75 |
| Hook-up Derrick Crane Block | 0.25 | 0.25 | 0.25 | 0.25 |
| Lift Jacket and Set in Water | 0.15 | 0.15 | 0.25 | 0.40 |
| Rotate and Position Jacket | 1.00 | 1.00 | 1.25 | 1.50 |
| Total Time Required | 9.40 | 9.65 | 10.40 | 11.15 |

The above operations include the following:

*Set Derrick Barge in Position:* Place derrick barge in correct location for setting jacket where required. With the use of a tug, haul anchors out and drop in desired location. Check anchors by pull method to be sure they are firmly secured to bottom.

*Set Material Barge in Position:* Position, secure and anchor material barge in correct location for derrick barge crane pick-up of jacket. If jacket is hauled to location on derrick barge, eliminate material barge from above.

*Break Loose Tie-Down:* Cut loose and remove all tie-down material used to secure jacket for shipment.

*Rig Jacket with Slings and Spreaders:* Place lifting slings and spreader bars, if required, to lifting pad eyes or lugs and prepare for lifting jacket.

*Hook-up Derrick Crane Block:* Position derrick crane block and fasten to lifting slings for lift.

*Lift Jacket and Set in Water:* With derrick crane, lift jacket from barge and set in water near its final location.

*Rotate and Position Jacket:* Rotate jacket as required and set in final position or resting place.

# PICK AND SET JACKETS
# Water Depths to 250 Feet

TOTAL SPREAD TIME REQUIRED IN HOURS

| Operation Description | TIME REQUIRED FOR FOLLOWING TONS | | | |
|---|---|---|---|---|
| | 0-200 | 201-300 | 301-400 | 401-500 |
| Set Derrick Barge in Position | 4.00 | 4.00 | 4.00 | 4.00 |
| Set Material Barge in Position | 1.00 | 1.00 | 1.00 | 1.00 |
| Break Loose Tie-Down | 2.50 | 2.75 | 3.00 | 3.25 |
| Rig Jacket with Slings and Spreaders | 0.50 | 0.50 | 0.65 | 0.75 |
| Hook-up Derrick Crane Block | 0.25 | 0.25 | 0.25 | 0.25 |
| Lift Jacket and Set in Water | 0.25 | 0.25 | 0.35 | 0.50 |
| Rotate and Position Jacket | 1.50 | 1.50 | 1.75 | 2.00 |
| Total Time Required | 10.00 | 10.25 | 11.00 | 11.75 |

The above operations include the following:

*Set Derrick Barge in Position:* Place derrick barge in correct location for setting jacket where required. With the use of a tug, haul anchors out and drop in desired location. Check anchors by pull method to be sure they are firmly secured to bottom.

*Set Material Barge in Position:* Position, secure and anchor material barge in correct location for derrick barge crane pick-up of jacket. If jacket is hauled to location on derrick barge, eliminate material barge from above.

*Break Loose Tie-Down:* Cut loose and remove all tie-down material used to secure jacket for shipment.

*Rig Jacket with Slings and Spreaders:* Place lifting slings and spreader bars, if required, to lifting pad eyes or lugs and prepare for lifting jacket.

*Hook-up Derrick Crane Block:* Position derrick crane block and fasten to lifting slings for lift.

*Lift Jacket and Set in Water:* With derrick crane, lift jacket from barge and set in water near its final location.

*Rotate and Position Jacket:* Rotate jacket as required and set in final position or resting place.

# PICK AND SET JACKETS
## Water Depths to 350 Feet

TOTAL SPREAD TIME REQUIRED IN HOURS

| Operation Description | TIME REQUIRED FOR FOLLOWING TONS | | | |
|---|---|---|---|---|
| | 0-200 | 201-300 | 301-400 | 401-500 |
| Set Derrick Barge in Position | 4.00 | 4.00 | 4.00 | 4.00 |
| Set Material Barge in Position | 1.00 | 1.00 | 1.00 | 1.00 |
| Break Loose Tie-Down | 2.50 | 2.75 | 3.00 | 3.25 |
| Rig Jacket with Slings and Spreaders | 0.50 | 0.50 | 0.65 | 0.75 |
| Hook-up Derrick Crane Block | 0.25 | 0.25 | 0.25 | 0.25 |
| Lift Jacket and Set in Water | 0.35 | 0.35 | 0.45 | 0.65 |
| Rotate and Position Jacket | 2.00 | 2.00 | 2.25 | 2.50 |
| Total Time Required | 10.60 | 10.85 | 11.60 | 12.40 |

The above operations include the following:

*Set Derrick Barge in Position:* Place derrick barge in correct location for setting jacket where required. With the use of a tug, haul anchors out and drop in desired location. Check anchors by pull method to be sure they are firmly secured to bottom.

*Set Material Barge in Position:* Position, secure and anchor material barge in correct location for derrick barge crane pick-up of jacket. If jacket is hauled to location on derrick barge, eliminate material barge from above.

*Break Loose Tie-Down:* Cut loose and remove all tie-down material used to secure jacket for shipment.

*Rig Jacket with Slings and Spreaders:* Place lifting slings and spreader bars, if required, to lifting pad eyes or lugs and prepare for lifting jacket.

*Hook-up Derrick Crane Block:* Position derrick crane block and fasten to lifting slings for lift.

*Lift Jacket and Set in Water:* With derrick crane, lift jacket from barge and set in water near its final location.

*Rotate and Position Jacket:* Rotate jacket as required and set in final position or resting place.

# LAUNCH AND SET JACKETS
## Water Depths to 150 Feet

TOTAL SPREAD TIME REQUIRED IN HOURS

| Operation Description | TIME REQUIRED FOR FOLLOWING TONS | | | |
|---|---|---|---|---|
| | 501-1000 | 1001-1500 | 1501-2000 | 2001-2500 |
| Set Derrick Barge in Position | 4.00 | 4.00 | 4.00 | 4.00 |
| Set Launch Barge in Position | 1.00 | 1.00 | 1.00 | 1.00 |
| Prepare Jacket and Launch Barge for Launch | 2.75 | 3.00 | 3.50 | 3.75 |
| Break Loose Tie-Down | 3.25 | 3.50 | 3.75 | 4.00 |
| Launch Jacket | 1.50 | 1.75 | 2.00 | 2.50 |
| Move Jacket to Location | 1.00 | 1.25 | 1.50 | 1.75 |
| Attach Crane Block and Flood Jacket Legs | 2.00 | 2.75 | 3.00 | 3.50 |
| Rotate and Position Jacket | 1.75 | 2.25 | 2.50 | 2.75 |
| Total Time Required | 17.25 | 19.50 | 21.25 | 23.25 |

The above operations include the following:

*Set Derrick Barge in Position:* Set derrick barge in correct location for setting jacket where required, with the use of a tug, haul anchors out and drop in desired location. Check anchors by pull method to be sure they are firmly secured on bottom.

*Set Launch Barge in Position:* Position, secure and anchor launch barge in line with derrick barge so derrick barge can assist with launch by the use of its winch lines and can pull and guide jacket to its position.

*Prepare Jacket and Launch Barge for Launch:* Rig jacket with slings and hold-back lines. Partially flood barge to allow jacket to slide off of launchways at an angle. Rig launch shoes to launch barge. Hook up launch barge and derrick barge winch lines to jacket for launch.

*Break Loose Tie-Down:* Cut loose and remove all tie-down material used to secure jacket for shipment.

*Launch Jacket:* With the use of the derrick barge winch line pulling and the launch barge winch line holding back, start jacket movement on launchways and launch.

*Move Jacket to Location:* With the use of tugs and hold-back lines and the derrick barge winch line, move jacket into position.

*Attach Crane Block and Flood Jacket Legs:* Attach the derrick barge crane main block to jacket slings for partial lifting of jacket.

*Rotate and Position Jacket:* Rotate jacket as may be required and set in final position or resting place on bottom.

# LAUNCH AND SET JACKETS
## Water Depths to 250 Feet

TOTAL SPREAD TIME REQUIRED IN HOURS

| Operation Description | TIME REQUIRED FOR FOLLOWING TONS | | | |
|---|---|---|---|---|
| | 501-1000 | 1001-1500 | 1501-2000 | 2001-2500 |
| Set Derrick Barge in Position | 4.00 | 4.00 | 4.00 | 4.00 |
| Set Launch Barge in Position | 1.00 | 1.00 | 1.00 | 1.00 |
| Prepare Jacket and Launch Barge for Launch | 2.75 | 3.00 | 3.50 | 3.75 |
| Break Loose Tie-Down | 3.25 | 3.50 | 3.75 | 4.00 |
| Launch Jacket | 1.50 | 1.75 | 2.00 | 2.50 |
| Move Jacket to Location | 1.00 | 1.25 | 1.50 | 1.75 |
| Attach Crane Block and Flood Jacket Legs | 2.00 | 2.75 | 3.00 | 3.50 |
| Rotate and Position Jacket | 2.25 | 2.75 | 3.00 | 3.25 |
| Total Time Required | 17.75 | 20.00 | 21.75 | 23.75 |

The above operations include the following:

*Set Derrick Barge in Position:* Set derrick barge in correct location for setting jacket where required, with the use of a tug, haul anchors out and drop in desired location. Check anchors by pull method to be sure they are firmly secured on bottom.

*Set Launch Barge in Position:* Position, secure and anchor launch barge in line with derrick barge so derrick barge can assist with launch by the use of its winch lines and can pull and guide jacket to its position.

*Prepare Jacket and Launch Barge for Launch:* Rig jacket with slings and hold-back lines. Partially flood barge to allow jacket to slide off of launchways at an angle. Rig launch shoes to launch barge. Hook up launch barge and derrick barge winch lines to jacket for launch.

*Break Loose Tie-Down:* Cut loose and remove all tie-down material used to secure jacket for shipment.

*Launch Jacket:* With the use of the derrick barge winch line pulling and the launch barge winch line holding back, start jacket movement on launchways and launch.

*Move Jacket to Location:* With the use of tugs and hold-back lines and the derrick barge winch line, move jacket into position.

*Attach Crane Block and Flood Jacket Legs:* Attach the derrick barge crane main block to jacket slings for partial lifting of jacket.

*Rotate and Position Jacket:* Rotate jacket as may be required and set in final position or resting place on bottom.

# LAUNCH AND SET JACKETS
## Water Depths to 350 Feet

TOTAL SPREAD TIME REQUIRED IN HOURS

| Operation Description | TIME REQUIRED FOR FOLLOWING TONS | | | |
|---|---|---|---|---|
| | 501-1000 | 1001-1500 | 1501-2000 | 2001-2500 |
| Set Derrick Barge in Position | 4.00 | 4.00 | 4.00 | 4.00 |
| Set Launch Barge in Position | 1.00 | 1.00 | 1.00 | 1.00 |
| Prepare Jacket and Launch Barge for Launch | 2.75 | 3.00 | 3.50 | 3.75 |
| Break Loose Tie-Down | 3.25 | 3.50 | 3.75 | 4.00 |
| Launch Jacket | 1.50 | 1.75 | 2.00 | 2.50 |
| Move Jacket to Location | 1.00 | 1.25 | 1.50 | 1.75 |
| Attach Crane Block and Flood Jacket Legs | 2.00 | 2.75 | 3.00 | 3.50 |
| Rotate and Position Jacket | 2.75 | 3.25 | 3.50 | 3.75 |
| Total Time Required | 18.25 | 20.50 | 22.25 | 24.25 |

The above operations include the following:

*Set Derrick Barge in Position:* Set derrick barge in correct location for setting jacket where required, with the use of a tug, haul anchors out and drop in desired location. Check anchors by pull method to be sure they are firmly secured on bottom.

*Set Launch Barge in Position:* Position, secure and anchor launch barge in line with derrick barge so derrick barge can assist with launch by the use of its winch lines and can pull and guide jacket to its position.

*Prepare Jacket and Launch Barge for Launch:* Rig jacket with slings and hold-back lines. Partially flood barge to allow jacket to slide off of launchways at an angle. Rig launch shoes to launch barge. Hook up launch barge and derrick barge winch lines to jacket for launch.

*Break Loose Tie-Down:* Cut loose and remove all tie-down material used to secure jacket for shipment.

*Launch Jacket:* With the use of the derrick barge winch line pulling and the launch barge winch line holding back, start jacket movement on launchways and launch.

*Move Jacket to Location:* With the use of tugs and hold-back lines and the derrick barge winch line, move jacket into position.

*Attach Crane Block and Flood Jacket Legs:* Attach the derrick barge crane main block to jacket slings for partial lifting of jacket.

*Rotate and Position Jacket:* Rotate jacket as may be required and set in final position or resting place on bottom.

# DRIVING TUBULAR PILES AND CONDUCTORS

HOURS REQUIRED PER LINEAR FOOT

| Nominal Size Inches | WALL THICKNESS IN INCHES | | | | | | |
|---|---|---|---|---|---|---|---|
| | 0.500 | 0.750 | 1.000 | 1.250 | 1.500 | 1.750 | 2.000 |
| 18 | 0.011 | 0.013 | 0.015 | 0.018 | 0.021 | 0.024 | 0.027 |
| 20 | 0.012 | 0.015 | 0.017 | 0.020 | 0.023 | 0.026 | 0.029 |
| 24 | 0.015 | 0.017 | 0.020 | 0.024 | 0.028 | 0.032 | 0.035 |
| 30 | 0.017 | 0.021 | 0.025 | 0.030 | 0.034 | 0.040 | 0.044 |
| 36 | 0.020 | 0.025 | 0.030 | 0.035 | 0.040 | 0.046 | 0.051 |
| 42 | 0.024 | 0.030 | 0.035 | 0.046 | 0.048 | 0.055 | 0.060 |
| 48 | 0.028 | 0.035 | 0.041 | 0.048 | 0.055 | 0.063 | 0.069 |
| 50 | 0.029 | 0.036 | 0.043 | 0.050 | 0.058 | 0.066 | 0.073 |
| 54 | 0.032 | 0.039 | 0.046 | 0.054 | 0.062 | 0.071 | 0.078 |

The above driving hours include:

1) Welding on and removal of cat eyes for hoisting and handling the pile or conductor.
2) The welding on and removal of stops necessary to control the pile or conductor for add-ons.
3) Picking up of piles or conductors and setting in driving leads.
4) Stabbing of piles or conductors.
5) Driving of piles or conductors.

Free fall and weight of hammer fall of pile or conductor are taken into consideration in the above time frames.

It is assumed that the proper size hammer and other equipment will be used for this operation.

See following tables for cutting and beveling, pre-heating, butt welds and radiographic inspection to be added to the above time frames if required.

# CUTTING AND BEVELING PILES AND CONDUCTORS

HOURS REQUIRED EACH

| Nominal Size Inches | WALL THICKNESS IN INCHES | | | | | | |
|---|---|---|---|---|---|---|---|
| | 0.500 | 0.750 | 1.000 | 1.250 | 1.500 | 1.750 | 2.000 |
| 18 | 0.255 | 0.285 | 0.300 | 0.337 | 0.382 | 0.450 | 0.472 |
| 20 | 0.285 | 0.327 | 0.360 | 0.397 | 0.420 | 0.487 | 0.555 |
| 24 | 0.420 | 0.475 | 0.487 | 0.500 | 0.550 | 0.607 | 0.682 |
| 30 | 0.545 | 0.610 | 0.635 | 0.670 | 0.695 | 0.715 | 0.745 |
| 36 | 0.695 | 0.745 | 0.785 | 0.815 | 0.845 | 0.870 | 0.900 |
| 42 | 0.930 | 0.965 | 0.990 | 1.025 | 1.080 | 1.120 | 1.655 |
| 48 | 1.230 | 1.270 | 1.310 | 1.350 | 1.390 | 1.445 | 1.480 |
| 50 | 1.290 | 1.287 | 1.390 | 1.435 | 1.477 | 1.550 | 1.573 |
| 54 | 1.380 | 1.428 | 1.474 | 1.518 | 1.563 | 1.625 | 1.665 |

# PRE-HEATING FOR BUTT WELDS

HOURS REQUIRED EACH

| Nominal Size Inches | WALL THICKNESS IN INCHES | | | | | | |
|---|---|---|---|---|---|---|---|
| | 0.500 | 0.750 | 1.000 | 1.250 | 1.500 | 1.750 | 2.000 |
| 18 | 0.045 | 0.065 | 0.078 | 0.091 | 0.102 | 0.110 | 0.136 |
| 20 | 0.052 | 0.072 | 0.092 | 0.109 | 0.122 | 0.139 | 0.161 |
| 24 | 0.060 | 0.086 | 0.094 | 0.111 | 0.125 | 0.140 | 0.166 |
| 30 | 0.098 | 0.105 | 0.130 | 0.154 | 0.179 | 0.199 | 0.229 |
| 36 | 0.107 | 0.118 | 0.138 | 0.172 | 0.205 | 0.225 | 0.257 |
| 42 | 0.114 | 0.129 | 0.158 | 0.189 | 0.253 | 0.270 | 0.307 |
| 48 | 0.120 | 0.132 | 0.173 | 0.214 | 0.254 | 0.295 | 0.336 |
| 50 | 0.128 | 0.141 | 0.183 | 0.228 | 0.271 | 0.314 | 0.358 |
| 54 | 0.135 | 0.149 | 0.193 | 0.241 | 0.286 | 0.332 | 0.378 |

The above cutting and beveling time frames include cutting and beveling of pile or conductor, in preparation for add-ons, using automatic beveling bands and cutting torch.

The above pre-heating for butt weld time frames include pre-heating of the pile or conductor tube for butt welds if required.

Cutting, beveling and pre-heating time frames should be added to the pile or conductor driving time frames if required.

# PICK AND SET DECKS

TOTAL SPREAD TIME REQUIRED

| Operation Description | TIME REQUIRED FOR FOLLOWING TONS | | | |
|---|---|---|---|---|
| | 0-200 | 200-300 | 300-400 | 400-500 |
| Set Derrick Barge in Position | 4.00 | 4.00 | 4.00 | 4.00 |
| Set Material Barge in Position | 1.00 | 1.00 | 1.00 | 1.00 |
| Break Loose Tie-Down | 2.00 | 2.00 | 2.50 | 2.50 |
| Rig Deck Section with Slings and Spreaders | 0.50 | 0.50 | 0.65 | 0.75 |
| Hook-Up Derrick Crane Block | 0.25 | 0.25 | 0.25 | 0.25 |
| Lift Deck Section and Set-To Jacket | 0.50 | 0.55 | 0.60 | 0.70 |
| Total Time Required | 8.25 | 8.30 | 9.00 | 9.20 |

The above operations include the following:

*Set Derrick Barge in Position:* Position, secure and anchor derrick barge in correct location for setting deck sections. If derrick barge is to remain in same position as was used to set jacket and piling, eliminate this time from the above spread.

*Set Material Barge in Position:* Position, secure and anchor material barge in correct location for derrick barge crane pick-up of deck sections. If material barge was previously positioned for off loading of other materials, delete this from above spread.

*Break Loose Tie-Down:* Cut loose and remove all tie-down material used to secure deck sections for shipment.

*Rig Deck Section with Slings and Spreaders:* Place lifting slings and spreader bars, if required, to lifting pad eyes or lugs and prepare for lifting deck sections.

*Hook-Up Derrick Crane Block:* Position derrick crane block and fasten to lifting slings for lift.

*Lift Deck Section and Set-To Jacket:* Lift and position deck section over jacket. Set deck stabbing guides into legs of jacket, release crane block, and remove slings and spreader bars.

See following tables for placement of shim plates and deck weld out time requirements to be added to the above spread time.

# SET SHIM PLATES

HOURS REQUIRED

| Operation Description | PLATE THICKNESS IN INCHES | | | | | | | |
|---|---|---|---|---|---|---|---|---|
| | 1 | 1-1/4 | 1-3/8 | 1-1/2 | 1-5/8 | 1-3/4 | 1-7/8 | 2 |
| Rig, Pick-up and Set—Each ..... | 0.17 | 0.17 | 0.20 | 0.20 | 0.20 | 0.23 | 0.23 | 0.23 |
| Weld Out—Per Linear Inch...... | 0.14 | 0.22 | 0.24 | 0.26 | 0.30 | 0.33 | 0.35 | 0.38 |

The above operations include the following:

*Rig, Pick-up and Set:* Fastening of slings, picking up of plate with derrick crane and setting plate in position.

*Weld Out:* Total weld out of shim plates in place. Only welders and welder helpers are required for this work. Balance of crew can be utilized on other operations.

# ERECT BOAT LANDINGS

HOURS REQUIRED

| Operation Description | TOTAL WEIGHT IN TONS | | |
|---|---|---|---|
| | 0-25 | 26-50 | 51-100 |
| Set Derrick Barge in Position .............. | 2.00 | 2.00 | 2.00 |
| Set Material Barge in Position ............. | 1.00 | 1.00 | 1.00 |
| Break Loose Tie-Down ..................... | 2.00 | 2.00 | 2.50 |
| Rig, Pick and Set-To Jacket ................ | 2.00 | 2.25 | 2.50 |
| Weld-out ................................ | 12.00 | 16.00 | 20.00 |
| Total Time Required ....................... | 19.00 | 23.25 | 28.00 |

The above time frames are for the complete installation and weld-out of prefabricated boat landings attaching to the jacket of platforms or other structures.

If derrick and material barges have been previously placed in position for other operations and repositioning is not necessary, eliminate the time frames for these operations from the above.

Use only that portion of the crew necessary for complete installation of the boat landing, leaving the balance of the crew to perform other duties.

# SET PLATFORM MISCELLANEOUS STEEL ITEMS AND WELD OUT

HOURS REQUIRED

| Operation Description | Unit | Hours Required |
|---|---|---|
| Weld Out Deck Sections | Ton | 0.25 |
| Erect Connecting Walkways | Ton | 0.10 |
| Erect Stairs | Ea. | 1.00 |
| Erect Ladders | Ea. | 1.00 |
| Erect Handrails | L.F. | 0.10 |
| Erect Grating | S.F. | 0.20 |
| Erect Floor Plate | S.F. | 0.25 |
| Erect Other Misc. Steel Items | Ton | 5.00 |

The above operations include the following:

*Weld Out Deck Sections:* Cutting and welding, as required, of beams, channels, plates, etc., for connecting; tying deck sections together and welding of deck stabbing guides to jacket. Above time frame is based on total tons of deck sections.

*Erect Connecting Walkways:* Rigging, picking, setting, aligning and fastening of walkways or pipeways between structures.

*Erect Stairs, Ladders and Handrails:* Rigging, picking, setting, aligning and securing in place.

*Erect Grating and Floor Plate:* Rigging, picking, setting, aligning and fastening in place.

*Erect Other Miscellaneous Steel Items:* Rigging, picking, cutting, welding, setting and aligning of other miscellaneous steel items not covered elsewhere.

Use only that portion of the crew necessary for complete installation of the individual item, leaving the balance of the crew to perform other duties.

# SET MISCELLANEOUS ITEMS ON PLATFORMS, JETTIES AND WHARFS

SPREAD HOURS REQUIRED

| Operation Description | Unit | Hours Required |
|---|---|---|
| Concrete Decks | Cy. | 10.00 |
| Wood Timber Decks | MBFM | 12.00 |
| Wood and Steel Fender Systems | S.F. | 0.10 |
| Barge Bumpers—Complete | S.F. | 0.20 |
| Tire Bumpers for Landings | Ea. | 0.10 |
| Heliport Decks | Ea. | 4.00 |
| Hooks and Capstans | Ea. | 1.50 |
| Quick-Release Hooks | Ea. | 1.00 |

The above operations include the following:

*Concrete Decks:* Placing and stripping forms, fabricating and placing reinforcing steel, placing embedded items and pouring and finishing concrete.

*Wood Timber Decks:* Cutting, placing and fastening timber for deck sections.

*Wood and Steel Fender Systems:* Placement of all steel supports and timber facings.

*Barge Bumpers Complete:* Placement of all steel supports and rubber bumpers.

*Tire Bumpers for Landings:* Placement of rubber tires to pre-installed structures.

*Heliport Decks:* Complete installation of all pre-fabricated structures and deck sections.

*Hooks and Capstans:* Installation of hooks and capstans.

*Quick-Release Hooks:* Installation of quick-release hooks.

# DRIVING STEEL SHEET AND "H" OR "I" BEAM PILING

HOURS REQUIRED PER UNITS LISTED

| Operation Description | Unit | Hours Required |
|---|---|---|
| Steel Sheet Piling: | — | — |
| Place and Drive | 100-Sq. Ft. | 10.00 |
| Pull or Remove | 100-Sq. Ft. | 5.00 |
| Cut-Off with Torch | 100-Lin. Ft. | 8.00 |
| Steel "H" or "I" Beams: | — | — |
| Haul and Unload at Site | Ton | 2.40 |
| Place and Drive | Lin. Ft. | 0.15 |
| Cut-Off with Torch | Each | 1.10 |

The above time frames include the following:

*Haul and Unload at Site:* Unloading at stockpile and loading, hauling and unloading at driving site if pile is to be placed with land rig. If piling is to be installed using marine equipment see section referring to load-out and tie-down.

*Place and Drive:* Hooking on, pulling into place and driving.

*Cut-Off with Torch:* Cutting off to desired elevation with cutting torch.

# ERECT STRUCTURAL AND MISCELLANEOUS STEEL FOR WHARFS AND JETTIES

## Structural Steel

SPREAD HOURS REQUIRED

| Operation Description | HOURS REQUIRED PER TON FOR FOLLOWING TONNAGE | | | |
|---|---|---|---|---|
| | To 500 | 500 To 1000 | 1000 To 5000 | Over 5000 |
| Unload and Shake Out .... | 3.0 | 3.0 | 2.5 | 2.0 |
| Erect, Plumb & Temp. Bolt. | 8.0 | 7.0 | 6.5 | 5.5 |
| Fastening By: Riveting .. | 6.5 | 6.0 | 5.0 | 4.5 |
| Welding ... | 4.0 | 3.5 | 2.5 | 2.0 |
| Bolting .... | 3.5 | 2.5 | 1.5 | 1.0 |

## Miscellaneous Steel

SPREAD HOURS REQUIRED

| Operation Description | Unit | Hours Required |
|---|---|---|
| 1½" Pipe Handrails ........................... | 100 Lin. Ft. | 23.0 |
| Angle Railing ................................ | Cwt. | 23.0 |
| Ladders ...................................... | Lin. Ft. | 0.3 |
| Stairs ....................................... | Cwt. | 5.5 |
| Curb Angle ................................... | Cwt. | 2.6 |
| Steel Curbing ................................ | Cwt. | 2.1 |
| Grating ...................................... | 100 Sq. Ft | 20.0 |
| Equipment Supports ........................... | Cwt. | 5.5 |
| Pipe Bridge Supports ......................... | Ton | 17.0 |
| Pipe Sleeves ................................. | Cwt. | 3.0 |
| Miscellaneous Pipe Hangers ................... | Cwt. | 3.0 |
| Wharf Tie-Back and Anchor Rods ............... | Cwt. | 2.0 |

Time requirements include rigging, picking, setting, aligning and fastening of the items as outlined above.

See other tables for concrete, painting and other miscellaneous item requirements.

# SINGLE-POINT MOORING TERMINAL

## General Notes

These general notes apply to the following operations and time requirement tables for the installation of single-point mooring terminals.

*Set Marker Bouys:* Layout and mark locations of all required items with buoys.

*Fasten Anchor Chains to Anchor Piles:* Weld pad eye attachment to anchor pile and fasten anchor chain to pad eye.

*Connect Pile Chaser to Anchor Pile:* If the top of the anchor pile is to be located on or near the bottom, the use of a removable pile chaser is recommended. The hours shown are for the time required to insert the chaser into the top of the pile and to secure same utilizing a portion of the anchor chain fastened to a pad eye located at the top of the chaser. This operation can be accomplished at the same time that the anchor chain is being attached to the pile.

*Install Anchor Pile:* The complete assembling and installation of the pile is included under this operation. See table on pile installation for driving time requirements.

*Remove Pile Chaser:* The cutting of the previously placed anchor chain from the chaser, the tie-in of the cut anchor chain to the balance of the anchor chain, and the removal of the chaser are included under this time frame.

*Install Terminal Buoy:* After all anchor piling or mooring legs have been installed, the terminal is to be towed or picked and set in its desired location.

*Pull and Tension Test Anchor Piles:* With the anchor chain shackled to the pile, the chain is laid out under tension as the barge is hauled back toward the system center. The time frame as shown will allow testing with a tension pull of up to 390 tons and is based on tons of chain.

*Connect Anchor Chains to Terminal Buoy:* All chains are hauled up through the terminal hawse pipes, fastened, and carefully tensioned. After tensioning, rubbing castings are fitted around the chains inside the hawse pipes.

# SINGLE-POINT MOORING TERMINAL

## General Notes (continued)

*Install Submerged Pipeline End Manifold:* Pick and set pipeline end manifold on the sea bottom in its desired location.

*Connect Pipeline to Manifold:* Connect submerged pipeline to manifold by bolting flanged end of pipeline to previously installed manifold flange. Time frame is based on size of line, in diameter inches, that is to be bolted up.

*Connect Pile Chaser to Manifold Anchor Pile:* This operation is similar to that outlined for "Connecting Pile Chaser to Anchor Pile." In the event that pile chaser is not used and pile is to be cut off under water increase the time frame by two times that as shown for this operation.

*Install Manifold Anchor Pile:* this operation is similar to that as outlined above for "Install Anchor Pile."

*Remove Manifold Anchor Pile Chaser:* This operation is similar to that outlined for "Remove Pile Chaser."

*Pin Anchor Pile to Manifold:* Cut each pile and insert and secure cross pins to prevent vertical movement of the pipeline end manifold.

*Install Underbuoy Hose System:* Preassemble hose strings and hydrostatically test to 225 psi. After successful test, string hose out in water. Time frame is based on 100 linear feet of hose or portion thereof.

*Manifold to Buoy Hose Hook-Up:* Attach one flanged end of the hose to flange on the underwater pipeline end manifold and secure the other end beneath the terminal. Time frame is based on size of line, in diameter inches, that is to be bolted up.

*Install Floating Hose System:* After successful test, as outlined above, the hose is to be launched and secured to the flange of the rotating cargo manifold of the buoy or terminal. Time frame includes hook-up and is based on 100 linear feet of hose or portion thereof.

*Install Float Ropes:* Launch floating ropes and attach to floating hose lines and terminal. Time frame is based on 100 lbs. or portion thereof.

*Install Hose End Lifting Gear:* Attach lifting gear to the loadout end of the floating hose.

*Install Winker Lights:* Install self-contained winker lights as required.

*Test System:* After installation is complete, hydrostatically test the entire system up to 225 psi. Time frame allowed is average for this type operation and may need adjusting depending on the scope of the job.

# SINGLE-POINT MOORING TERMINAL

TOTAL SPREAD TIME REQUIRED

| Operation Description | Unit | Hours Required |
|---|---|---|
| Set Market Buoys | Ea. | 2.0 |
| Fasten Anchor Chain to Anchor Pile | Ea. | 2.0 |
| Connect Pile Chaser to Anchor Pile | Ea. | 0.5 |
| Install Anchor Piles | Lf. | See Notes |
| Remove Pile Chasers | Ea. | 1.0 |
| Install Terminal Buoy | Ton | 0.2 |
| Pull and Tension Test Anchor Piles | Ton | 1.4 |
| Connect Anchor Chains to Terminal Buoy | Ea. | 2.0 |
| Install Submerged Pipeline End Manifold | Ton | 0.4 |
| Connect Pipeline to Manifold | Dia. Inch | 0.6 |
| Connect Pile Chaser to Manifold Anchor Pile | Ea. | 0.5 |
| Install Manifold Anchor Pile | Lf. | See Notes |
| Remove Manifold Anchor Pile Chaser | Ea. | 1.5 |
| Pin Anchor Pile to Manifold | Ea. | 2.0 |
| Install Underbuoy Hose System | C. LF. | 12.0 |
| Manifold to Buoy Hose Hook-Up | Dia. Inch | 0.6 |
| Install Floating Hose System | C. LF. | 6.0 |
| Install Float Ropes | C. LB. | 0.6 |
| Install Hose End Lifting Gear | Ea. | 2.0 |
| Install Winker Lights | Ea. | 1.0 |
| Test System | Ea. | 24.0 |

Reasonable time has been included in the above time frames for positioning construction equipment for work, break loose of tie-down and rigging of equipment or item to be installed.

Some of the operations listed above can be performed simultaneously. This must be left to the judgement of the estimator. Where this occurs, simply ignore the time requirements for minor operation involved.

See tables on pile installation for time required to drive piling.

See General Notes on preceding pages for descriptive information regarding the individual operations listed above.

# PLATFORM OR STRUCTURE INSTALLATION

**Derrick Barge Labor Crew**

| Personnel Description | Number of Men |
|---|---|
| Superintendent or Captain | 1 |
| Assistant Superintendent | 1 |
| Barge Foreman | 2 |
| Welder Foreman | 2 |
| Rigger/Pile Butt Foreman | 2 |
| Anchor Foreman | 2 |
| Leader Man | 4 |
| Crane Operator—To 600 Ton | 2 |
| Crane Operator—100 Ton | 2 |
| Hoist Operator | 4 |
| Anchor Hoist Operator | 4 |
| Fireman/Oiler | 4 |
| Structural Welders | 16 |
| Welder Helpers | 16 |
| Riggers/Pile Butts | 12 |
| Laborers | 10 |
| Ironworkers | 10 |
| X-Ray Technician | 2 |
| X-Ray Helpers | 4 |
| Clerk | 2 |
| Radio Operator/Technician | 1 |
| Assistant Clerk/Radio Operator | 1 |
| Field Engineer | 1 |
| Crewmen | 10 |
| Total Crew | 115 |

The above labor crew is for the supervision and installation of offshore platforms and structures and should be ample for the work intended as outlined in the preceding tables.

See other labor crews in this section that must be added for a complete working spread.

# OFFSHORE CONSTRUCTION

## Derrick Barge Equipment Spread

| Equipment Description | Number of Units |
|---|---|
| Barge with Revolving 600 Ton Crane | 1 |
| Crawler Crane—100 Ton | 1 |
| Anchor Handling Winches | 8 |
| Anchors—12,000 Pounds Each ± | 8 |
| 1½" Anchor Wire Rope—3,000 Lin. Ft. Ea. | 8 |
| Generators—250 KW. | 4 |
| Air Compressor—600 CFM | 2 |
| Steam Boiler—900 HP, 200 PSI | 1 |
| Pile Driving Leads | 1 |
| Pile Hammers (Various Sizes) | 4 |
| Steam Hose—400 Feet | 2 |
| Welding Machines—400 Amp | 20 |
| X-Ray Equipment (Complete Set-up) | 1 |
| Water Pumps—4" | 2 |
| Water Pumps—6" | 4 |
| Short Wave Radio System | 1 |

For the installation of offshore platforms, structures, mooring terminals and equipment installation, the derrick barge should consist of or be equipped with the above.

See other equipment spreads in this section that must be added to the above for a complete working spread.

If jetting is required for pile installation, add the following jetting equipment to the above.

| Equipment Description | Number of Units |
|---|---|
| Jetting Pumps—2,000 GPM @ 1,000 PSI | 4 |
| Dredge Pump—20 Inch | 1 |
| Suction and Jetting Ladder | 1 |
| Jetting Sled | 1 |

# SINGLE-POINT MOORING TERMINAL INSTALLATION

**Derrick Barge Labor Crew**

| Personnel Description | Number of Men |
|---|---|
| Superintendent or Captain | 1 |
| Assistant Superintendent | 1 |
| Barge Foreman | 2 |
| Pipe Welder Foreman | 2 |
| Rigger/Pile Butt Foreman | 2 |
| Anchor Foreman | 2 |
| Leaderman | 2 |
| Crane Operator—600 Ton | 2 |
| Crane Operator—100 Ton | 2 |
| Hoist Operator | 2 |
| Anchor Hoist Operator | 2 |
| Fireman/Oiler | 2 |
| Pipe Welders | 12 |
| Welder Helpers | 12 |
| Riggers/Pile Butts | 12 |
| Iron Workers | 4 |
| Electricians | 2 |
| Laborers | 4 |
| X-Ray Technician | 2 |
| Clerk | 1 |
| Radio Operator/Technician | 1 |
| Assistant Clerk/Radio Operator | 1 |
| Field Engineer | 1 |
| Crewman | 6 |
| Total Crew | 80 |

The above labor crew is for the supervision and installation of single-point mooring terminals and should be ample to accomplish the work as outlined on the total spread time required table for this type installation.

See other tables in this section that must be added for a complete working spread.

# DERRICK BARGE MAINTENANCE OPERATION

**Labor Crew**

| Personnel Description | Number of Men |
|---|---|
| Chief Engineer | 1 |
| Shift Engineer | 2 |
| Machinist | 2 |
| Electrician | 2 |
| Engineer | 2 |
| Oiler | 2 |
| First Aid Man | 1 |
| Total Crew | 12 |

The above labor crew is for maintaining the derrick barge and assuring that all its mechanical and electrical inclusions function properly and that emergency first aid is available.

# DERRICK OR QUARTER BARGE QUARTERING AND CATERING

**Labor Crew**

| Personnel Description | Number of Men |
|---|---|
| Chief Cook | 2 |
| Second Cook | 2 |
| Chief Stewart | 1 |
| Mess Boy | 2 |
| Janitor | 4 |
| Laundry Man | 2 |
| Total Crew | 13 |

Duties of the above labor crew consist of preparing and serving meals, house cleaning and maintaining laundry facilities for the derrick barge personnel.

These duties can be performed either on the derrick barge or on a special quartering barge, depending on the available facilities.

# TUG SPREAD

## Labor Crew

| Personnel Description | Number of Men |
|---|---|
| Captain | 1 |
| First Mate | 1 |
| Chief Engineer | 1 |
| First Engineer | 1 |
| Seaman | 4 |
| Oiler | 1 |
| Cook | 1 |
| Total Crew | 10 |

## Equipment Spread

| Equipment Description | Number of Units |
|---|---|
| Tug | 1 |

The above labor crew and equipment spread are for maintaining and operating a tug of the correct size to handle and tow the derrick barge and material barges and to handle, where and when required, the anchors of the aforementioned barges.

The size, horsepower and rating of the tug will be governed by the equipment it is to handle and the waters it is to be towing or working in.

For small tugs with limited capacity, working relatively close to shore and in relatively calm waters, the above labor crew can be adjusted downward.

# CREW BOAT SPREAD

## Labor Crew

| Personnel Description | Number of Men |
|---|---|
| Captain | 1 |
| Mate/Engineer | 1 |
| Seaman | 1 |
| Total Crew | 3 |

## Equipment Spread

| Equipment Description | Number of Units |
|---|---|
| Crew Boat—65′ ± | 1 |

The above labor crew and equipment spread are for hauling personnel, round trip as may be required, between shore and installation or work site.

When and where feasible, the spread can also be utilized for hauling miscellaneous light supplies such as welding rod and food items.

The size of the total crews and location of the project will govern the size and number of crew boats required.

# DIVER SPREAD

## Labor Crew

| Personnel Description | Number of Men |
|---|---|
| Diver Supervisor | 1 |
| Divers | 4 |
| Diver Tenders | 4 |
| Total Crew | 9 |

## Equipment Spread

| Equipment Description | Number of Units |
|---|---|
| Decompression Chamber | 2 |
| High Compression Compressor | 2 |
| Diving Gear | 4 |
| Oxygen Quads | 4 |
| Lot of Small Hand Tools | 1 |

The above labor crew and equipment spread are ample to perform all underwater duties such as inspection of bottom, jacket and grouting and underwater cut-offs.

Adjustments to the above labor crew and equipment spread may be required depending on water depth and quantity and type of operations to be performed.

# WORK OR SUPPLY BOAT SPREAD

## Operating Labor Crew

| Personnel Description | Number of Men |
|---|---|
| Captain | 1 |
| First Mate | 1 |
| Chief Engineer | 1 |
| First Engineer | 1 |
| Electrician | 1 |
| Oiler | 1 |
| Seaman | 4 |
| Cook | 1 |
| Total Crew | 11 |

## Equipment Spread

| Equipment Description | Number of Units |
|---|---|
| Work Boat | 1 |

The above labor crew and equipment spread are for maintaining and operating a work or supply boat required to haul supplies or assist in certain special operations such as support for survey crews, etc.

The size, horsepower and rating of the boat will be governed by the duties it is to perform and the waters it is to be performing in.

For small boats with limited capacity, working close to shore in relatively calm waters, the above labor crew can be adjusted downward.

# SURVEY SPREAD

## Labor Crew

| Personnel Description | Number of Men |
|---|---|
| Chief Survey Engineer | 1 |
| Survey Engineers | 3 |
| Helpers | 6 |
| Total Crew | 10 |

## Equipment Spread

| Equipment Description | Number of Units |
|---|---|
| Work boat (Fully Manned) | 1 |
| Survey Equipment | 1 |

The purpose of the above labor crew and equipment spread are for surveying and laying out the proposed facilities.

Length of time required for this operation will be governed by the scope of the project.

See page 201 for fully-manned boat requirements.

# SANDBLAST AND PAINT LABOR AND EQUIPMENT SPREAD

## Labor Crew

| Personnel Description | Number of Men |
|---|---|
| Paint Foreman | 1 |
| Blasters and Painters | 2 |
| Pot Man/Helper | 3 |
| Operator | 1 |
| Total Crew | 7 |

## Equipment Spread

| Equipment Description | Number of Units |
|---|---|
| Air Compressor—600 CFM | 1 |
| Sand Pot and Spray Equip. | 1 |
| Air Hose—100 Feet | 1 |
| Sand Blast Hose—200 Feet | 1 |
| Lot of Small Tools | 1 |

The above labor crew and equipment spread are for cleaning (by blasting) and touch-up painting of various structural and process items after installation.

If possible, the above labor crew should be substituted from other craft aboard the derrick barge. If not feasible these personnel should be brought aboard at the tail end of the construction and only for the length of time required for this operation.

# DRILLING OPERATION

## Labor Crew

| Personnel Description | Number of Men |
|---|---|
| Superintendent | 1 |
| Foreman | 2 |
| Drillers | 8 |
| Driller Helpers | 8 |
| Compressor Operator | 2 |
| Pump Operator | 2 |
| Total Crew | 23 |

## Equipment Spread

| Equipment Description | Number of Units |
|---|---|
| Drill Rig with Support Equipment | 1 |

The above foreman through pump operator should be minimum personnel required for a two-shift operation for each drill rig required or used.

Many types of drill rigs can be used. Some of the determining factors are type of soil and size of hole to be drilled, working space and load-bearing capacities, and amount and quantity of excavated material to be removed.

All necessary supporting equipment, such as power unit, rotary table, drive train, derrick mast, hoist, traveling carriage, support structure, insert collars, controls, hydraulic system, drill pipe, power sub and swivel, drill bits under reamer, auger head, mud tanks and pumps, water pumps, and tongs, hose and other drilling tools, should be furnished with each drilling rig.

Where extensive drilling is to be performed, a work barge with tug will probably be required to support the drilling operation.

# GROUT INSTALLATION

## Labor Crew

| Personnel Description | Number of Men |
|---|---|
| Foreman | 2 |
| Cement Plant Operator | 2 |
| Cement Plant Helper | 12 |
| Grout Pump Operator | 2 |
| Pump Operator | 2 |
| Compressor Operator | 2 |
| Total Crew | 22 |

## Equipment Spread

| Equipment Description | Unit | Number of Units |
|---|---|---|
| Work Barge—120′ × 40′ × 8′ | Each | 1 |
| Deck House and Tool Shed | Each | 1 |
| Submersible Water Pump—5 HP. | Each | 2 |
| Generator—5 KW. | Each | 2 |
| Grout Mix Equipment | Each | 1 |
| Grout Pump | Each | 1 |
| Surge Tank | Each | 3 |
| Water Tank | Each | 1 |
| Air Compressor—300 CFM. | Each | 1 |
| Piping | Lot | 1 |

Above labor crew and equipment spread are ample for a two-shift operation where a major quantity of grout is to be placed.

The total yardage and conditions will dictate the size and amount of equipment required.

Where minor yardage of grout is involved, a small skid-mounted unit is advisable for this operation. Should this be the case, and if ample space is available on the derrick barge for the grout equipment and materials, the above equipment should be eliminated and the labor crew adjusted downward to fit the needs.

If the grout unit is furnished on a rental basis by an organization specializing in this type of operation, they will usually quote a day rate plus expenses for a supervisor. If required, this must be added.

# PRODUCTIVITY AND PERCENTAGE EFFICIENCY FACTORS

To correctly apply the time frames in the preceding tables, consideration must be given to the effects of labor productivity and to wind, waves and currents.

It must be recognized that, when performing work at sea, it is usually wise to complement the project with the full labor and equipment spreads required to accomplish the work involved, although all the operations cannot be performed simultaneously (exceptions to this rule are if the project is close enough to shore and ample skilled craftsmen are available and can easily be shuttled to and from the project site or if craftsmen skilled in more than one of the required trades are available).

It should also be recognized that when the project has a full complement of crew and equipment, some of the operations involved can be performed concurrently and no additional time need be allowed for the labor and equipment spreads. For example, once the piling operations are underway, a portion of the cutting and beveling, pre-heating, welding and X-raying can be performed simultaneously or painting touch-up can be accomplished as other work progresses.

To assist in the installation estimation of offshore structures, we offer the following labor, wind, wave and current efficiency percentages to apply to the preceding time frames under varied conditions. An example of the application of these factors is included. Following this example we have inserted some actual itemized time spent tables for various structures, showing those items of work that were accomplished concurrently, all of which were installed under very similar conditions as outlined in the example. The actual labor crew and equipment spread for the installation of these structures are also included.

Productivity factors to be applied to offshore structures and underwater piling installations for labor are as follows:

LABOR PRODUCTIVITY FACTORS

| Type of Productivity | Percentage Range |
|---|---|
| Excellent | 0.901 through 1.000 |
| Very Good | 0.801 through 0.900 |
| Average | 0.601 through 0.800 |
| Low | 0.401 through 0.600 |
| Very Low | 0.001 through 0.400 |

The following production elements should be given consideration for the application of a labor productivity percentage.

1. General Economy
2. Project Supervision
3. Labor Conditions
4. Job Conditions
5. Equipment
6. Weather

For full description of above elements and example of obtaining labor productivity percentage, see pages xi-xiv.

Productivity factors to be applied for offshore platforms and underwater piling in regard to efficient operations under certain average wind speeds in miles per hour are as follows:

WIND PRODUCTIVITY FACTORS

| Description | Wind Miles Per Hour | Percent Efficiency |
|---|---|---|
| Calm | 0-1 | 100 |
| Light Air | 1-3 | 100 |
| Slight Breeze | 4-7 | 95 |
| Gentle Breeze | 8-12 | 90 |
| Moderate Breeze | 13-18 | 75 |
| Fresh Breeze | 19-24 | 50 |
| Strong Breeze | 25-31 | 30 |

The above wind descriptions and speeds have been taken from Beaufort's Wind Scale, as modified for steam.

Percentage efficiency as listed above may fluctuate slightly, depending on type of sea-going equipment to be used, presence of wind breakers when necessary, experience of personnel, etc.

Above percentage efficiency is based on actual time and motion studies made from platform and underwater pile installations under average wind speeds as listed above.

Productivity factors to be applied for offshore platform installation in regard to efficient operations under certain average current speeds in feet per second are as follows:

CURRENT PRODUCTIVITY FACTORS

| Average Total Current in Feet Per Second | Percent Efficiency |
|---|---|
| 0.0 to 0.5 | 100 |
| 0.5 to 1.0 | 97 |
| 1.0 to 2.0 | 95 |
| 2.0 to 2.5 | 90 |
| 2.5 to 3.0 | 85 |
| 3.0 to 3.5 | 78 |
| 3.5 to 4.0 | 70 |
| 4.0 to 5.0 | 65 |

Currents are of sufficient magnitude that they can and do influence installation operations for platform jackets and underwater piling.

Above average total current represents total average of tidal current, density current and wind drift current.

Above percent efficiency is based on actual time and motion studies made from platform jacket and piling installations under the average current speeds as listed.

Productivity factors to be applied for various pieces of sea-going equipment in regard to efficient operations under certain wave conditions are as follows:

WAVE PRODUCTIVITY FACTORS

| Equipment and Type of Operations | WAVE HEIGHT IN FEET AND PERCENTAGE EFFICIENCY FOR: | | | | | |
|---|---|---|---|---|---|---|
| | Safe Efficient Operations | | Marginal Operations | | Dangerous and/or Inefficient Operations | |
| | Wave Height | Percent Efficiency | Wave Height | Percent Efficiency | Wave Height | Percent Efficiency |
| Deep Sea Tug: | | | | | | |
| Towing Derrick Barge | 0-4 | 100-70 | 4-6 | 70-50 | 6 or greater | 50-20 |
| Towing Material Barge | 0-4 | 100-70 | 4-6 | 70-50 | 6 or greater | 50-20 |
| Working Derrick Barge | 0-2 | 100-70 | 2-3 | 70-40 | 3 or greater | 40-10 |
| Working Material Barge | 0-2 | 100-70 | 2-3 | 70-40 | 3 or greater | 40-10 |
| Crew Boats (60 to 90 Feet Long): | | | | | | |
| Underway | 0-8 | 100-80 | 8-15 | 80-40 | 15 or greater | 40-10 |
| Loading or Unloading Crews | 0-3 | 100-70 | 3-5 | 70-50 | 5 or greater | 50-20 |
| Derrick Barge: | | | | | | |
| Small Barge—Underway | 0-2 | 100-70 | 2-3 | 70-50 | 3 or greater | 50-20 |
| Large Barge—Underway | 0-3 | 100-70 | 3-5 | 70-50 | 5 or greater | 50-20 |
| Small Barge—Platform Building | 0-2 | 100-70 | 2-3 | 70-40 | 3 or greater | 40-10 |
| Large Barge—Platform Building | 0-3 | 100-70 | 3-4 | 70-40 | 4 or greater | 40-10 |
| Small Barge—Buoy Laying | 0-2 | 100-70 | 2-3 | 70-40 | 3 or greater | 40-10 |
| Ship-Mounted Derrick: | | | | | | |
| Platform Building | 0-4 | 100-70 | 4-6 | 70-50 | 6 or greater | 50-20 |

Wave heights used are those of the average maximum waves for the work intended.

Wave height is defined as the average height of the highest one-third of the wave passing a stationary location during a given observation period.

Wave height limits given above are not rigid but will vary to some extent with locality, local wind conditions, experience of personnel, etc.

The above wave limits will fluctuate slightly, depending on the position of the vessel relative to the direction of wave approach, and it is recommended that the vessel be oriented with its bow into prevailing seas while working.

# EXAMPLE—APPLICATION OF FACTORS

The following example is to show how to apply the labor, wind, wave and current factor tables in this section to estimate the actual time requirements for a given project.

We assume that we are going to install a drilling and production platform offshore and that a detailed estimate has been made utilizing the various preceding tabels and that we estimate an overall installation time of 15 days at one hundred percent productivity for all elements.

Next we consider labor, wind, wave and currents, which have a direct bearing on the actual productivity we can expect to obtain.

From the labor analyzation table (page 206) and the example of labor productivity (pages xi-xiv) it has been determined that our labor productivity efficiency is equal to sixty-five percent, and for the purpose of this example we will assume this to be correct.

Wind, wave and current effects can be determined from examination and study of meteorological-oceanographic factors, and other information affecting the area and time in which our project is to be constructed. After review of this information, we determine the average wind speed, wave height and current speed and select from the preceding tables the conditions and efficiency percentages that apply. We then total the condition percentages and divide by the total number of conditions to obtain an average percentage efficiency to be applied against the total predetermined days. For out example we assume the following:

| Item | Conditions | Percent Efficiency |
|---|---|---|
| Average Wind .......... | 4-7 Miles Per Hour | 95 |
| Average Wave Height ... | 2 Feet Average Height | 91 |
| Average Current ........ | 2 Feet Per Second | 95 |
| Total .................. | — | 281 |

$$\frac{281}{3} = 94 \text{ Percent Efficiency Average}$$

It should be noted that both labor and wind, wave and current conditions individually affect the length of time required for the project.

# EXAMPLE—APPLICATION OF FACTORS (continued)

Now that we have established an efficiency percentage that we expect to obtain from labor and one we expect to obtain from wind, wave and current conditions, we arrive at multiple factors to apply to the predetermined platform installation days. This is accomplished by dividing a total efficiency percentage of 100 by the actual determined percent, or as follows:

Labor: $\frac{100}{65} = 1.54$ Multiple Factor

Wind, Wave and Current: $\frac{100}{94} = 1.06$ Multiple Factor

To arrive at the actual days required for our project, the following formula applies:

Number of Predetermined Days × Labor Multiple Factor × Wind, Wave and Current Multiple Factor = Actual Estimated Days Required

or

15 Days × 1.54 × 1.06 = 24.50 or 25 Days Required

Simply by applying the proper labor crews and equipment spreads and their respective costs for the total adjusted days, a total direct cost for the platform can easily be obtained.

# ACTUAL WORKING CREW FOR INSTALLATION OF THREE-, FOUR-, SIX-, NINE- AND TEN-PILE STRUCTURES

**Labor Crew**

| Personnel Description | NUMBER OF MEN FOR | |
|---|---|---|
| | 3-, 4- and 6-Pile Structures | 9- and 10-Pile Structures |
| Superintendent or Captain | 1 | 1 |
| Assistant Superintendent | 1 | 1 |
| Barge Foreman | 2 | 2 |
| Welder Foreman | 2 | 2 |
| Rigger/Pile Butt Foreman | 2 | 2 |
| Anchor Foreman | 2 | 2 |
| Leaderman | 2 | 4 |
| Crane Operator—To 600 Ton | 2 | 2 |
| Crane Operator—100 Ton | 2 | 2 |
| Hoist Operator | 2 | 4 |
| Anchor Hoist Operator | 4 | 4 |
| Fireman/Oiler | 2 | 4 |
| Structural Welders | 15 | 18 |
| Welder Helpers | 15 | 18 |
| Riggers/Pile Butts | 10 | 12 |
| Laborers | 8 | 10 |
| Ironworkers | 8 | 10 |
| X-Ray Technician | 2 | 2 |
| X-Ray Helpers | 2 | 4 |
| Clerk | 2 | 2 |
| Radio Operator/Technician | 1 | 1 |
| Assistant Clerk/Radio Operator | 1 | 1 |
| Field Engineer | 1 | 1 |
| Crewmen | 8 | 10 |
| Total Crew | 97 | 119 |

The above labor crews as outlined were used for the installation of the following structures and dolphins.

In addition to the above, the following crews, as previously outlined, were used and added to the above (see equipment spread for number of tug crews required):

1) Derrick Barge Maintenance Crew
2) Quartering and Catering Crew
3) Tug Crew
4) Crew Boat Crew
5) Diver Crew
6) Survey Crew
7) Sand Blast and Paint Crew
8) Grouting Crew

# ACTUAL EQUIPMENT SPREADS FOR INSTALLATION OF THREE-, FOUR-, SIX-, NINE- AND TEN-PILE STRUCTURES

**Equipment Spread**

| Equipment Description | NUMBER OF UNITS FOR | |
|---|---|---|
| | Pick and Set Structures | Launched Structures |
| Derrick Barge—300′ × 100′ × 21′ | 1 | 1 |
| Cargo Barge—240′ × 72′ × 15′ | 1 | 0 |
| Cargo Barge—306′ × 68′ × 15.7′ | 1 | 1 |
| Launch Barge—240′ × 72′ × 15′ | 0 | 1 |
| Grout Barge—150′ × 40′ × 12′ | 1 | 1 |
| Tug—2,900 HP. | 1 | 1 |
| Tug—1,600 HP. | 1 | 2 |
| Crew Boat—65′ | 1 | 1 |

The above derrick barge is fully equipped, as outlined under the derrick barge equipment spread table. All necessary survey, diving and paint equipment is also included.

Cargo barges are equipped with winches, cables and anchors.

Launch barge is equipped with the necessary launchways and rocker arms, ballasting and deballasting systems and winches, cables and anchors.

Grout barge is equipped with all necessary grout equipment and barge anchors.

# ACTUAL TIME FRAME FOR INSTALLING 12-WELL DRILLING PLATFORM

## 165-ft. Water Depth, 6-Pile Structure, 150-ft. Penetration

CREW HOURS BY OPERATION

| Operation Description | Unit | Quantity | OPERATION TIME | |
|---|---|---|---|---|
| | | | Required | Concurrent |
| Set Derrick Barge in Position | Ea. | 1 | 7 | — |
| Set Material Barge in Position | Ea. | 2 | 3 | — |
| Break Loose, Pick and Set Jacket in Water | Ton | 447 | 9 | — |
| Rotate and Position Jacket | Ton | 447 | 3 | — |
| Stab and Drive Piles and Conductors | LF. | 6120 | 209 | — |
| Cut and Bevel Piles and Conductors | Ea. | 36 | 27 | 27 |
| Pre-Heat Piles and Conductors | Ea. | 36 | 6 | 6 |
| Butt Weld Piles and Conductors | Ea. | 18 | 232 | 232 |
| X-Ray Welds | Ea. | 18 | — | 16 |
| Set Grout Seals | Ea. | 6 | 10 | — |
| Grout Legs | Cy. | 108 | 3 | — |
| Break Loose, Pick and Set Deck | Ton | 165 | 5 | — |
| Set Shim Plates | Ea. | 24 | 160 | — |
| Break Loose, Pick and Set Boat Landing | Ton | 25 | — | 7 |
| Set Stairs and Handrails | Lot | 1 | — | 49 |
| Weld Out Deck and Boat Landing | Ton | 190 | — | 77 |
| Paint Touch-Up | Lot | 1 | — | 12 |
| Total Hours Required | | | 674 | |

Platform consists of:

Piles—6 each, 30″ O.D. × 340′ long, average ¾″ wall.
Conductors—12 Each, 30″ O.D. × 340′ long, average ⅝″ wall.

Jacket Size at Bottom—75.75′ × 105.75′
Jacket Size at Top—31.25′ × 61.25′
Jacket Height—185.5′
Jacket Weight—447 short tons.

Cellar Deck Size—60′ × 30′
Main Deck Size—80′ × 40′
Double Deck Weight—165 short tons.

Boat Landing Weight—25 short tons.

Miscellaneous—240 linear feet of loose handrails, 4 stairs and 2 ladders.

Weight of decks includes grating floor covering, and cellar deck handrail.

# ACTUAL TIME FRAME FOR INSTALLING 4-WELL DRILLING PLATFORM

## 165-ft. Water Depth, 4-Pile Structure, 150-ft. Penetration

CREW HOURS BY OPERATION

| Operation Description | Unit | Quantity | OPERATION TIME | |
|---|---|---|---|---|
| | | | Required | Concurrent |
| Set Derrick Barge in Position | Ea. | 1 | 7 | — |
| Set Material Barge in Position | Ea. | 2 | 3 | — |
| Break Loose, Pick and Set Jacket in Water | Ton | 282 | 6 | — |
| Rotate and Position Jacket | Ton | 282 | 2 | — |
| Stab and Drive Piles and Conductors | LF. | 2720 | 93 | — |
| Cut and Bevel Piles and Conductors | Ea. | 16 | 8 | 8 |
| Pre-Heat Piles and Conductors | Ea. | 32 | 3 | 2 |
| Butt Weld Piles and Conductors | Ea. | 16 | 106 | 100 |
| X-Ray Welds | Ea. | 16 | — | 7 |
| Set Grout Seals | Ea. | 4 | 7 | — |
| Grout Legs | Cy. | 62 | 2 | — |
| Break Loose, Pick and Set Deck | Ton | 125 | 5 | — |
| Set Shim Plates | Ea. | 16 | 106 | — |
| Set Stairs and Handrails | Lot | 1 | — | 41 |
| Weld Out Deck and Boat Landing | Ton | 150 | — | 61 |
| Paint Touch-Up | Lot | 1 | — | 8 |
| Total Hours Required | | | 348 | |

Platform consists of:

Piles—4 each, 30″ O.D. × 340′ long, average ¾″ wall.
Conductors—4 each, 30″ O.D. × 340′ long, average ¾″ wall.

Jacket Size at Bottom—75.75′ × 85.75′
Jacket Size at Top—31.25′ × 41.25′
Jacket Height—180.5′
Jacket Weight—282 short tons (includes boat landing).

Cellar Deck Size—30′ × 30′
Main Deck Size—60′ × 40′
Double Deck and Boat Landing Weight—125 short tons.

Miscellaneous—200 linear feet of loose handrails, 2 stairs and 3 ladders.

Weight of deck includes grating floor covering, and cellar deck handrail.

# ACTUAL TIME FRAME FOR INSTALLING PRODUCTION PLATFORM

## 165-ft. Water Depth, 4-Pile Structure, 150-ft. Penetration

CREW HOURS BY OPERATION

| Operation Description | Unit | Quantity | OPERATION TIME | |
|---|---|---|---|---|
| | | | Required | Concurrent |
| Set Derrick Barge in Position | Ea. | 1 | 7 | — |
| Set Material Barge in Position | Ea. | 2 | 3 | — |
| Break Loose, Pick and Set Jacket in Water | Ton | 322 | 7 | — |
| Rotate and Position Jacket | Ton | 322 | 3 | — |
| Stab and Drive Piles | L.F. | 1360 | 47 | — |
| Cut and Bevel Piles | Ea. | 8 | 4 | 4 |
| Pre-Heat Piles | Ea. | 16 | 2 | 1 |
| Butt Weld Piles | Ea. | 8 | 53 | 50 |
| X-Ray Welds | Ea. | 8 | — | 4 |
| Set Grout Seals | Ea. | 4 | 7 | — |
| Grout Legs | Cy. | 62 | 2 | — |
| Break Loose, Pick and Set Deck | Ton | 165 | 5 | — |
| Set Shim Plates | Ea. | 16 | 106 | — |
| Set Stairs and Handrails | Lot | 1 | — | 46 |
| Set Disposal Conductor | Ea. | 1 | — | 4 |
| Weld Out Deck and Boat Landing | Ton | 190 | 27 | 50 |
| Set Heliport Deck | Ea. | 1 | 7 | — |
| Paint Touch-Up | Lot | 1 | — | 10 |
| Total Hours Required | | | 280 | |

Platform consists of:

Piles—4 each, 30″ O.D. × 340′ long, average ¾″ wall.
Disposal Conductors—1 each, 30″ O.D. × 65′ long.

Jacket Size at Bottom—85.75′ × 85.75′
Jacket Size at Top—41.25′ × 41.25′
Jacket Height—180.5′
Jacket Weight—322 short tons (includes boat landing).

Cellar Deck Size—40′ × 40′
Main Deck Size—55′ × 50′
Double Deck—165 short tons.

Octagon Heliport Deck Size—40′ × 40′ Overall

Miscellaneous—210 linear feet of loose handrail, 4 stairs and 3 ladders.

Weight of deck includes grating floor covering and cellar deck handrail.

# ACTUAL TIME FRAME FOR INSTALLING GATHERING AND PRODUCTION PLATFORM
## 165-ft. Water Depth, 6-Pile Structure, 150-ft. Penetration

CREW HOURS BY OPERATION

| Operation Description | Unit | Quantity | OPERATION TIME | |
|---|---|---|---|---|
| | | | Required | Concurrent |
| Set Derrick Barge in Position | Ea. | 1 | 7 | — |
| Set Material Barge in Position | Ea. | 2 | 3 | — |
| Break Loose, Launch and Float Jacket | Ton | 536 | 17 | — |
| Flood, Sink, Rotate and Position Jacket | Ton | 536 | 4 | — |
| Stab and Drive Piles | LF. | 2040 | 70 | — |
| Cut and Bevel Piles | Ea. | 18 | 10 | 8 |
| Pre-Heat Piles | Ea. | 24 | 2 | 2 |
| Butt Weld Piles | Ea. | 12 | 115 | 40 |
| X-Ray Welds | Ea. | 12 | — | 5 |
| Set Grout Seals | Ea. | 6 | 10 | — |
| Grout Legs | Cy. | 108 | 3 | — |
| Break Loose, Pick and Set Deck | Ton | 280 | 5 | — |
| Set Shim Plates | Ea. | 24 | 160 | — |
| Break Loose, Pick and Set Boat Landing | Ton | 25 | — | 7 |
| Set Stairs and Handrails | Lot | 1 | — | 38 |
| Weld Out Deck and Boat Landing | Ton | 305 | — | 76 |
| Paint Touch-Up | Lot | 1 | — | 10 |
| Total Hours Required | | | 406 | |

Platform consists of:

Piles—6 each, 30″ O.D. × 340′ long, average ¾″ wall.

Jacket Size at Bottom—125.75′ × 75.75′
Jacket Size at Top—81.25′ × 31.25′
Jacket Height—188.0′
Jacket Weight—536 short tons.

Cellar Deck Size—80′ × 30′
Main Deck Size—100′ × 50′
Double Deck Weight—280 short tons.

Boat Landing Weight—25 short tons.

Miscellaneous—300 linear feet of loose handrails, 4 stairs and 4 ladders. Jacket weight includes a grated walkway and handrails. Weight of deck includes grating floor covering, cellar deck handrails, and minor steel equipment foundations or supports.

# ACTUAL TIME FRAME FOR INSTALLING POWER AND PUMP PLATFORM

## 165-ft. Water Depth, 9-Pile Structure, 150-ft. Penetration

CREW HOURS BY OPERATION

| Operation Description | Unit | Quantity | OPERATION TIME | |
|---|---|---|---|---|
| | | | Required | Concurrent |
| Set Derrick Barge in Position | Ea. | 1 | 7 | — |
| Set Material Barge in Position | Ea. | 2 | 3 | — |
| Break Loose, Launch and Float Jacket | Ton | 712 | 17 | — |
| Flood, Sink, Rotate and Position Jacket | Ton | 712 | 4 | — |
| Stab and Drive Piles | LF. | 3060 | 105 | — |
| Cut and Bevel Piles | Ea. | 27 | 15 | 12 |
| Pre-Heat Piles | Ea. | 36 | 3 | 3 |
| Butt Weld Piles | Ea. | 18 | 140 | 92 |
| X-Ray Welds | Ea. | 18 | — | 8 |
| Set Grout Seals | Ea. | 9 | 15 | — |
| Grout Legs | Cy. | 162 | 4 | — |
| Break Loose, Pick and Set Deck | Ton | 540 | — | 7 |
| Set Shim Plates | Ea. | 36 | 239 | — |
| Break Loose, Pick and Set Boat Landing | Ton | 30 | — | 8 |
| Set Stairs and Handrails | Lot | 1 | — | 47 |
| Weld Out Deck and Boat Landing | Ton | 570 | — | 143 |
| Paint Touch-Up | Lot | 1 | — | 12 |
| Total Hours Required | | | 552 | |

Platform consists of:

Piles—9 each, 30″ O.D. × 340′ long, average ¾″ wall.

Jacket Size at Bottom—125.75′ × 115.75′
Jacket Size at Top—81.25′ × 71.00′
Jacket Height—190.5′
Jacket Weight—712 short tons.

Cellar Deck Size—80′ × 70′
Main Deck Size—106.75′ × 80′
Double Deck Weight—540 short tons.

Boat Landing Weight—30 tons.

Miscellaneous—374 linear feet of loose handrails, 6 stairs and 4 ladders. Jacket weight includes a grated walkway and handrails. Double deck is in three separate sections and includes grating floor covering, cellar deck handrails, and minor steel equipment foundations or supports.

# ACTUAL TIME FRAME FOR INSTALLING HIGH-PRESSURE FLARE STRUCTURE

## 165-ft. Water Depth, 3-Pile Structure, 150-ft. Penetration

CREW HOURS BY OPERATION

| Operation Description | Unit | Quantity | OPERATION TIME | |
|---|---|---|---|---|
| | | | Required | Concurrent |
| Set Derrick Barge in Position | Ea. | 1 | 7 | — |
| Set Material Barge in Position | Ea. | 1 | 2 | — |
| Break Loose, Pick and Set Jacket in Water | Ton | 111 | 6 | — |
| Rotate and Position Jacket | Ton | 111 | 2 | — |
| Stab and Drive Piles | LF. | 870 | 24 | — |
| Cut and Bevel Piles | Ea. | 6 | 3 | 2 |
| Pre-Heat Piles | Ea. | 12 | 1 | 1 |
| Butt Weld Piles | Ea. | 6 | 46 | 23 |
| X-Ray Welds | Ea. | 6 | — | 2 |
| Set Grout Seals | Ea. | 3 | 5 | — |
| Grout Legs | Cy. | 58 | 2 | — |
| Break Loose, Pick and Set Deck | Ton | 21 | 5 | — |
| Set Shim Plates | Ea. | 12 | 80 | — |
| Set Flare Stack Tower | Ea. | 1 | 2 | — |
| Set Flare Stack | Ea. | 1 | 2 | — |
| Weld Out Deck, Tower and Stack | Ton | 135 | — | 34 |
| Paint Touch-Up | Lot | 1 | — | 6 |
| Total Hours Required | | | 187 | |

Platform consists of:

Piles—3 each, 24″ O.D. × 290′ long, average ¾″ wall.

Tripot Jacket Size at Bottom—66.75′ between legs.
Tripot Jacket Size at Top—15.42′ between legs.
Tripot Jacket Height—185.5′
Tripot Jacket Weight—111 short tons.

Single Deck Size—14′ × 14′
Single Deck Weight—21 short tons.

Flare Stack Tower Height—132′

Flare Stack Height—150′

# ACTUAL TIME FRAME FOR INSTALLING LOW PRESSURE FLARE STRUCTURE

## 165-ft. Water Depth, 4-Pile Structure, 150-ft. Penetration

CREW HOURS BY OPERATION

| Operation Description | Unit | Quantity | OPERATION TIME | |
|---|---|---|---|---|
| | | | Required | Concurrent |
| Set Derrick Barge in Position | Ea. | 1 | 7 | — |
| Set Material Barge in Position | Ea. | 1 | 2 | — |
| Break Loose, Pick and Set Jacket in Water | Ton | 162 | 6 | — |
| Rotate and Position Jacket | Ton | 162 | 2 | — |
| Stab and Drive Piles | LF. | 1160 | 32 | — |
| Cut and Bevel Piles | Ea. | 8 | 4 | 2 |
| Pre-Heat Piles | Ea. | 16 | 1 | 1 |
| Butt Weld Piles | Ea. | 8 | 50 | 43 |
| X-Ray Welds | Ea. | 8 | — | 3 |
| Set Grout Seals | Ea. | 4 | 4 | — |
| Grout Legs | Cy. | 77 | 2 | — |
| Break Loose, Pick and Set Deck | Ton | 21 | 5 | — |
| Set Shim Plates | Ea. | 16 | 106 | — |
| Set Flare Stack Tower | Ea. | 1 | 2 | — |
| Set Flare Stack | Ea. | 1 | 2 | — |
| Weld Out Deck, Tower and Stack | Ton | 186 | — | 76 |
| Paint Touch-Up | Lot | 1 | — | 10 |
| Total Hours Required | | | 225 | |

Platform consists of:

Piles—4 each, 24" O.D. × 290′ long, average ¾" wall.

Jacket Size at Bottom—66.29′ × 66.29′
Jacket Size at Top—15.42′ × 15.42′
Jacket Height—185.5′
Jacket Weight—162 short tons.

Single Deck Size—14′ × 14′
Single Deck Weight—21 short tons.

Flare Stack Tower Height—132′

Flare Stack Height—150′

Miscellaneous—Deck section includes grating floor covering and pre-installed handrails.

# ACTUAL TIME FRAME FOR INSTALLING MAIN BREASTING DOLPHIN

## 100-ft. Water Depth, 10-Pile Structure, 150-ft. Penetration

CREW HOURS BY OPERATION

| Operation Description | Unit | Quantity | OPERATION TIME | |
|---|---|---|---|---|
| | | | Required | Concurrent |
| Set Derrick Barge in Position | Ea. | 1 | 7 | — |
| Set Material Barge in Position | Ea. | 2 | 3 | — |
| Break Loose, Launch and Float Jacket | Ton | 644 | 17 | — |
| Flood, Sink, Rotate and Position Jacket | Ton | 644 | 3 | — |
| Stab and Drive Piles | LF. | 2850 | 163 | — |
| Cut and Bevel Piles | Ea. | 10 | 10 | 6 |
| Pre-Heat Piles | Ea. | 20 | 3 | 2 |
| Butt Weld Piles | Ea. | 10 | 319 | — |
| X-Ray Welds | Ea. | 10 | — | 14 |
| Set Grout Seals | Ea. | 10 | 16 | — |
| Grout Legs | Cy. | 445 | 12 | — |
| Break Loose, Pick and Set Deck | Ton | 62 | 5 | — |
| Set Shim Plates | Ea. | 40 | 275 | — |
| Set Stairs and Handrails | Lot | 1 | — | 24 |
| Install Capstans | Ea. | 3 | — | 5 |
| Install Quick Release Hooks | Ea. | 4 | — | 4 |
| Weld Out Deck | Ton | 62 | — | 16 |
| Install Fenders | SF. | 1760 | 87 | 200 |
| Paint Touch-Up | Lot | 1 | — | 12 |
| Total Hours Required | | | 920 | |

Dolphin consists of:

Piles—10 each, 42" O.D. × 285′ long, average 1" wall.

Jacket Size at Top and Bottom—39.83′ × 69.00′
Jacket Height—124.67′
Jacket Weight—644 short tons.

Lower Deck Size—69.00′ × 40.00′
Upper Deck Size—46.00′ × 39.83′
Double Deck Weight—62 short tons.

Fenders—2 each, 40′ long × 22′ high.

Miscellaneous—126 linear feet of loose handrail and 2 stairs.

Weight of decks includes grating floor covering and lower deck handrails.

# ACTUAL TIME FRAME FOR INSTALLING SECONDARY BREASTING DOLPHIN

## 100-ft. Water Depth, 10-Pile Structure, 150-ft. Penetration

CREW HOURS BY OPERATION

| Operation Description | Unit | Quantity | OPERATION TIME | |
|---|---|---|---|---|
| | | | Required | Concurrent |
| Set Derrick Barge in Position | Ea. | 1 | 7 | — |
| Set Material Barge in Position | Ea. | 2 | 3 | — |
| Break Loose, Launch and Float Jacket | Ton | 496 | 17 | — |
| Flood, Sink, Rotate and Position Jacket | Ton | 496 | 3 | — |
| Stab and Drive Piles | LF. | 2850 | 116 | — |
| Cut and Bevel Piles | Ea. | 10 | 6 | 6 |
| Pre-Heat Piles | Ea. | 20 | 2 | 2 |
| Butt Weld Piles | Ea. | 10 | 100 | 91 |
| X-Ray Welds | Ea. | 10 | — | 9 |
| Set Grout Seals | Ea. | 10 | 16 | — |
| Grout Legs | Cy. | 353 | 6 | — |
| Break Loose, Pick and Set Deck | Ton | 81 | 5 | — |
| Set Shim Plates | Ea. | 40 | 266 | — |
| Set Stairs and Handrails | Lot | 1 | — | 24 |
| Install Capstans | Ea. | 3 | — | 5 |
| Install Quick Release Hooks | Ea. | 4 | — | 4 |
| Weld Out Deck | Ton | 81 | — | 33 |
| Install Fenders | SF. | 1760 | 100 | 187 |
| Set Heliport Deck | Ea. | 1 | 4 | — |
| Paint Touch-Up | Lot | 1 | — | 12 |
| Total Hours Required | | | 651 | |

Dolphin consists of:

Piles—10 each, 36″ O.D. × 285′ long, average ¾″ wall.

Jacket Size at Top and Bottom—39.83′ × 69.00′
Jacket Height—124.67′
Jacket Weight—496 short tons.

Lower Deck Size—40′ × 69′
Upper Deck Size—39.83′ × 46′
Double Deck Weight—81 short tons.

Fenders—2 each, 40′ long × 22′

Miscellaneous—126 linear feet of loose handrail and 2 stairs.

Weight of decks includes grating floor covering, and lower deck handrails.

# ACTUAL TIME FRAME FOR INSTALLING AUXILIARY BREASTING DOLPHIN

## 100-ft. Water Depth, 10-Pile Structure, 150-ft. Penetration

CREW HOURS BY OPERATION

| Operation Description | Unit | Quantity | OPERATION TIME | |
|---|---|---|---|---|
| | | | Required | Concurrent |
| Set Derrick Barge in Position | Ea. | 1 | 7 | — |
| Set Material Barge in Position | Ea. | 2 | 3 | — |
| Break Loose, Launch and Float Jacket | Ton | 599 | 17 | — |
| Flood, Sink, Rotate and Position Jacket | Ton | 559 | 3 | — |
| Stab and Drive Piles | LF. | 2850 | 116 | — |
| Cut and Bevel Piles | Ea. | 10 | 6 | 6 |
| Pre-Heat Piles | Ea. | 20 | 2 | 2 |
| Butt Weld Piles | Ea. | 10 | 100 | 91 |
| X-Ray Welds | Ea. | 10 | — | 9 |
| Set Grout Seals | Ea. | 10 | 16 | — |
| Grout Legs | Cy. | 353 | 6 | — |
| Break Loose, Pick and Set Deck | Ton | 60 | 5 | — |
| Set Shim Plates | Ea. | 40 | 266 | — |
| Set Stairs and Handrails | Lot | 1 | — | 24 |
| Install Capstans | Ea. | 3 | — | 5 |
| Install Quick Release Hooks | Ea. | 4 | — | 4 |
| Weld Out Deck | Ton | 60 | — | 15 |
| Install Fenders | SF. | 880 | 43 | 100 |
| Paint Touch-Up | Lot | 1 | — | 10 |
| Total Hours Required | | | 590 | |

Dolphin consists of:

Piles—10 each, 36″ O.D. × 285′ long, average ¾″ wall.

Jacket Size at Top and Bottom—39.83′ × 69.00′
Jacket Height—124.67′
Jacket Weight—599 short tons.

Lower Deck Size—69.00′ × 40.00′
Upper Deck Size—46.00′ × 39.83′
Double Deck Weight—60 short tons.

Fenders—2 each, 20′ long × 22′ high.

Miscellaneous—126 linear feet of loose handrail and 2 stairs.

Weight of decks includes grating floor covering and lower deck handrails.

# ACTUAL TIME FRAME FOR INSTALLING LOADING PLATFORM
## 100-ft. Water Depth, 6-Pile Structure, 150-ft. Penetration

CREW HOURS BY OPERATION

| Operation Description | Unit | Quantity | OPERATION TIME | |
|---|---|---|---|---|
| | | | Required | Concurrent |
| Set Derrick Barge in Position | Ea. | 1 | 7 | — |
| Set Material Barge in Position | Ea. | 2 | 3 | — |
| Break Loose, Launch and Float Jacket | Ton | 549 | 17 | — |
| Flood, Sink, Rotate and Position Jacket | Ton | 549 | 3 | — |
| Stab and Drive Piles | LF. | 1710 | 59 | — |
| Cut and Bevel Piles | Ea. | 6 | 3 | 3 |
| Pre-Heat Piles | Ea. | 12 | 1 | 1 |
| Butt Weld Piles | Ea. | 6 | 50 | 27 |
| X-Ray Welds | Ea. | 6 | — | 3 |
| Set Grout Seals | Ea. | 6 | 6 | — |
| Grout Legs | Cy. | 131 | 2 | — |
| Break Loose, Pick and Set Deck | Ton | 386 | 5 | — |
| Set Shim Plates | Ea. | 24 | 160 | — |
| Break Loose, Pick and Set Boat Landings | Ton | 60 | — | 8 |
| Set Stairs and Handrails | Lot | 1 | — | 37 |
| Weld Out Deck and Boat Landings | Ton | 446 | 82 | 100 |
| Break Loose, Pick and Set Quarter Bldg. | Ea. | 1 | — | 5 |
| Paint Touch-Up | Lot | 1 | — | 20 |
| Total Time Required | | | 398 | |

Platform consists of:

Piles—6 each, 30″ O.D. × 285′ long, average ¾″ wall.

Jacket Size at Bottom—109.50′ × 73.58′
Jacket Size at Top—81.25′ × 51.00′
Jacket Weight—549 short tons.

Cellar Deck Size—80.00′ × 50.00′
Main Deck Size—104.00′ × 74.00′
Double Deck Weight—386 short tons.

Boat Landing Weight—30 short tons each (two required).

Quarter Building—3-story, 40.00′ × 30.00′ overall size.

Miscellaneous—182 linear feet of loose handrail and 5 stairs.

Weight of deck includes grating cover on cellar deck, ⅜″ checkered plate covering on main deck and cellar deck handrails.

# ACTUAL TIME FRAME FOR INSTALLING BRIDGES

CREW HOURS BY OPERATION

| Operation Description | Unit | Quantity | OPERATION TIME | |
|---|---|---|---|---|
| | | | Required | Concurrent |
| Set Derrick Barge in Position | Ea. | 1 | 14 | — |
| Set Material Barge in Position | Ea. | 1 | 4 | — |
| Break Loose, Pick and Set Bridges | Ton | 103 | 10 | — |
| Weld Out Bridge Sections | Ton | 103 | 3 | — |
| Paint Touch-Up | Lot | 1 | — | 10 |
| Total Hours Required | | | 31 | |

Bridges consist of:

Bridges—1 each, 100′ long, weighing 19 short tons.
Bridges—1 each, 330′ feet long, weighing 84 short tons.

Miscellaneous—Bridges are three pole pipe structures with grating walks and handrails.

# Section Seven

# OFFSHORE INSTALLATION OF EQUIPMENT AND APPURTENANCES

Time frames, in manhours required for the installation of equipment, pre-fabricated piping spool tie-ins, electrical installations, and instrumentation installations on offshore structures, are included under this section. We have also included labor crew and equipment spread tables which we feel will meet maximum requirements for these installations.

The labor listed in the tables is for a double crew, each working a 12-hour shift, complementing a 24-hour day, utilizing the equipment spreads full time.

In applying the hours included in the following time frame tables, consideration must be given to items that will affect the overall productivity that one might expect to obtain on a project of this nature. Once the derrick barge is anchored and secured in its position, ocean currents will have little or no effect on productivity. However, wind, waves and labor will have a decided effect and should be given consideration in obtaining an overall productivity factor to apply against the hours.

When performing work at sea, it is usually wise to complement the project with the full labor crew and equipment spread that will be required for the work involved. Once the work is underway many operations can be performed concurrently and no time need be assessed for these operations. For example, once certain pieces of equipment are installed, tie-in of piping, electrical and instrumentation installations can be performed while other pieces of equipment are being set.

The estimator should prepare a step-by-step time requirement table, segregating the major items that will require the spread from those that can be performed concurrently, and assess the crew and spread only for the time required to perform the major operations. For examples of these time frames, see Section Six, "Offshore Structures," pages 213 through 224.

For efficiency percentages to apply against labor, wind and waves and an example of their application, see Section Six, "Offshore Structures," pages 206 through 210.

# SKID-MOUNTED PACKAGED UNITS

HOURS REQUIRED FOR WEIGHT UNITS LISTED

| Packaged Units (Weight in Tons) | Hours Required Per Ton | Minimum Total Hours Required |
|---|---|---|
| 0- 5 | 15.0 | 30 |
| 6- 10 | 8.0 | 50 |
| 11- 20 | 6.0 | 75 |
| 21- 30 | 5.0 | 125 |
| 31- 40 | 5.0 | 175 |
| 41- 50 | 4.5 | 200 |
| 51- 60 | 4.5 | 250 |
| 61-100 | 4.0 | 275 |
| 101-200 | 3.0 | 325 |
| 201-300 | 1.5 | 360 |
| 301-400 | 1.2 | 375 |
| 401-450 | 1.1 | 450 |

Hours required include break-loose of tie-down, rigging, picking, setting, aligning and fastening in position of pre-assembled skid-mounted equipment or building packages.

Above minimum total hours required are the minimum that should be used per package that falls within the listed weight spreads.

Time requirements do not include piping, electrical or instrumentation tie-ins to other packages or pieces of equipment. See other tables for these time frames.

The above hours will suffice for the installation of any type of pre-assembled skid-mounted equipment package or building package.

# SETTING INDIVIDUAL TANKS, VESSELS AND HEAT EXCHANGERS

HOURS REQUIRED FOR WEIGHT UNITS LISTED

| Item Description (Weight in Tons) | Hours Required Per Ton | Minimum Total Hours Required |
|---|---|---|
| Tanks or Vessels: | | |
| 0- 5 | 10.0 | 20 |
| 6- 10 | 8.0 | 60 |
| 11- 20 | 7.0 | 90 |
| 21- 30 | 5.0 | 110 |
| 31- 40 | 4.0 | 130 |
| 41- 50 | 4.0 | 170 |
| 51- 60 | 3.0 | 180 |
| 61-100 | 3.0 | 200 |
| Heat Exchangers: | | |
| Double Pipe—All Weights | 6.0 | 25 |
| Shell and Tube—To 3 Tons | 20.0 | 25 |
| Shell and Tube—From 3 Tons | 5.0 | 60 |
| Drums and Pots—All Weights | 7.0 | 25 |

Hours required include break-loose of tie-down, rigging, picking, setting, aligning, and fastening in position of individual piece of equipment.

Above minimum total hours required are the minimum that should be used for each piece of equipment within the listed weight spreads.

Time requirements do not include piping, electrical or instrumentation tie-ins to systems or equipment. See other tables for these time frames.

The above hours are ample for the installation of equipment such as separators, scrubbers, dryers, air receivers, degassing boots, etc.

# SETTING PUMPS AND COMPRESSORS

HOURS REQUIRED FOR HORSEPOWERS LISTED

| Item Description<br>Horsepower Rating | Hours Required<br>Per H.P. | Minimum Total<br>Hours Required |
|---|---|---|
| Pumps with Drives: | | |
| 0- 25 HP | 5.00 | 25 |
| 26- 75 HP | 3.00 | 125 |
| 76- 125 HP | 2.00 | 225 |
| 126- 500 HP | 1.00 | 250 |
| 501-7500 HP | 0.50 | 500 |
| Chemical Injection Pumps | 25.00 | 25 |
| Compressors with Drives: | | |
| Centrifugal 0- 100 HP | 3.00 | 100 |
| Centrifugal 101- 500 HP | 2.00 | 300 |
| Centrifugal 501-1500 HP | 1.00 | 1000 |
| Centrifugal 1501-2500 HP | 0.75 | 1500 |
| Centrifugal 2501-3500 HP | 0.65 | 1875 |
| Centrifugal 3501-5000 HP | 0.50 | 2275 |
| Instru. Air Compressors | — | 120 |

Hours required include break-loose of tie-down, rigging, picking, setting, aligning, and fastening in position of pumps or centrifugal compressors with drives.

Above minimum total hours required are the minimum for each piece of equipment within the listed horsepower spreads.

For reciprocating compressors, increase the above hours 15 percent.

Time requirements do not include piping, electrical or instrumentation tie-ins to systems or equipment. See other tables for these time frames.

# SETTING DIESEL GENERATORS

HOURS REQUIRED EACH

| Diesel Generator Rating | Hours Required |
|---|---|
| 35 KW—1800 RPM—60 Cycle—3 Phase | 25 |
| 45 KW—1800 RPM—60 Cycle—3 Phase | 30 |
| 60 KW—1800 RPM—60 Cycle—3 Phase | 34 |
| 90 KW—1800 RPM—60 Cycle—3 Phase | 38 |
| 125 KW—1800 RPM—60 Hertz—3 Phase | 40 |
| 200 KW—1800 RPM—60 Hertz—3 Phase | 50 |
| 250 KW—1800 RPM—60 Hertz—3 Phase | 55 |

Above hours include break-loose of tie-down, rigging, picking, setting, aligning, and fastening in position of self-contained diesel generators of the capacities listed.

Time requirements do not include piping, electrical or instrumentation tie-ins. See other tables for these time frames.

# MISCELLANEOUS EQUIPMENT AND ITEMS

HOURS REQUIRED PER UNITS LISTED

| Item Description | Unit | Hours Required |
|---|---|---|
| Jib Cranes—2- To 5-Ton Capacity | Ea. | 4 |
| Pedestal Mounted Crane—15-Ton Capacity | Ea. | 12 |
| Pedestal Mounted Crane—30-Ton Capacity | Ea. | 18 |
| High-Pressure Flare—60 MMSCFD | Ea. | 14 |
| Low-Pressure Flare—7.5 MMSCFD | Ea. | 16 |
| Low-Pressure Flare—8-Inch Flame Arrestor | Ea. | 16 |
| Flare Tips | Ea. | 30 |

Above hours are average required for break-loose of tie-down, rigging, picking, setting and aligning of the items as listed.

It is assumed that all the above items will be pre-assembled, in so far as practical, prior to load-out and tie-down for shipment.

# MARINE LOADING ARMS

## General Notes and Description

*DCMA—Double Counterweighted Marine Arms:* The DCMA arm has a system of pantograph sheaves, cables, and two sets of counterweights. Larger arms may be supplemented with hydraulic power if desired. Power is provided through hydraulics operable from a master dock control station, control tower, or a remote control unit.

*PCMA—Pantograph Counterweighted Marine Arms:* The PCMA arm has a pantograph sheave arrangement, cables, and a single counterweight. The basic PCMA design provides for hydraulic operation of the outboard leg only. Power for the arm is provided in the same manner as that listed above for DCMA arms.

*HMLA—Hydraulic Marine Loading Arms:* The HMLA arm is hydraulically-powered in all movements. Its hydraulics, counterweights, and power equipment are mounted within the support frame and access structure. Power for this arm is accomplished in much the same manner as that shown for DCMA arms above.

*CMA—Counterweighted Marine Arms:* The CMA arm is a counterbalanced marine arm basically designed for manual operation.

The assembling hours listed in the following tables include the receiving and shaking-out of all arm materials and the assembling of the various components insofar as practical prior to installation.

The assembling time frames allow for installation of hydraulic power components for all DCMA arms 12-inches in diameter or greater and 51 feet long or greater and for installation of hydraulic power components for PCMA and HMLA arms as outlined above and required.

The installation hours listed in the table on pages 232 and 233 include time for the completion of arm assembly, arm installation and testing.

The hours listed in these tables do not include time for the installation of master control stands, control towers or remote control centers. If required, time must be added for these installations.

Actual time required for the assembling and installation of loading arms will be governed by the size of crew and the hours worked per day. Simply by dividing the total hours required (as outlined in the following tables) by the productive crew hours determined, a total time frame can easily be established.

# MARINE LOADING ARMS

**Assembling Only**

HOURS REQUIRED EACH ARM

| Arm Nominal Size Inches | OVERALL LENGTH OF ARM IN FEET AND HOURS REQUIRED TO ASSEMBLE | | | | | | | | |
|---|---|---|---|---|---|---|---|---|---|
| | 0-20 | 21-30 | 31-40 | 41-50 | 51-60 | 61-70 | 71-80 | 81-90 | 91-100 |
| DCMA Arms: | | | | | | | | | |
| 4 | 25 | 32 | 40 | 51 | 65 | 83 | 105 | 126 | 160 |
| 6 | 31 | 40 | 50 | 63 | 79 | 100 | 126 | 159 | 200 |
| 8 | 44 | 56 | 69 | 86 | 107 | 133 | 164 | 200 | 242 |
| 10 | 57 | 72 | 86 | 103 | 124 | 149 | 178 | 213 | 255 |
| 12 | 72 | 91 | 107 | 126 | 178 | 210 | 248 | 293 | 346 |
| 14 | 80 | 102 | 117 | 135 | 186 | 214 | 256 | 302 | 356 |
| 16 | 93 | 118 | 132 | 148 | 198 | 222 | 266 | 319 | 383 |
| 18 | 102 | 130 | 143 | 156 | 206 | 227 | 272 | 326 | 391 |
| 20 | 117 | 150 | 164 | 179 | 234 | 257 | 283 | 339 | 407 |
| 24 | 140 | 178 | 189 | 200 | 254 | 279 | 307 | 362 | 427 |
| PCMA Arms: | | | | | | | | | |
| 6 | 42 | 53 | 67 | 85 | 108 | — | — | — | — |
| 8 | 59 | 75 | 95 | 121 | 154 | — | — | — | — |
| 10 | 77 | 98 | 124 | 157 | 199 | — | — | — | — |
| 12 | 97 | 123 | 156 | 198 | 250 | — | — | — | — |
| HMLA Arms: | | | | | | | | | |
| 6 | 56 | 71 | 90 | 114 | 145 | — | — | — | — |
| 8 | 79 | 100 | 127 | 161 | 204 | — | — | — | — |
| 10 | 103 | 130 | 165 | 210 | 267 | — | — | — | — |
| 12 | 130 | 165 | 210 | 267 | 339 | — | — | — | — |
| CMA Arms: | | | | | | | | | |
| 4 | 22 | 28 | 36 | 46 | 58 | 67 | 77 | 89 | 104 |
| 6 | 28 | 36 | 46 | 58 | 74 | 81 | 93 | 107 | 124 |
| 8 | 40 | 51 | 65 | 83 | 105 | 116 | 129 | 146 | 168 |
| 10 | 51 | 65 | 83 | 105 | 133 | 145 | 160 | 179 | 202 |
| 12 | 65 | 83 | 105 | 133 | 170 | 183 | 199 | 219 | 245 |
| 14 | 72 | 91 | 115 | 141 | 172 | 187 | 202 | 224 | 251 |
| 16 | 84 | 107 | 135 | 166 | 180 | 194 | 211 | 234 | 257 |
| 18 | 92 | 117 | 147 | 181 | 195 | 207 | 226 | 246 | 268 |
| 20 | 105 | 133 | 168 | 207 | 219 | 232 | 253 | 271 | 293 |
| 24 | 126 | 160 | 202 | 212 | 223 | 239 | 258 | 279 | 296 |

The above hours are for assembling of arms. See page 231 for general notes and description of arms; time requirements for installation of arms are on page 233.

# MARINE LOADING ARMS

**Erection Only**

HOURS REQUIRED EACH ARM

| Arm Nominal Size Inches | OVERALL LENGTH OF ARM IN FEET AND HOURS REQUIRED TO ERECT | | | | | | | | |
|---|---|---|---|---|---|---|---|---|---|
| | 0-20 | 21-30 | 31-40 | 41-50 | 51-60 | 61-70 | 71-80 | 81-90 | 91-100 |
| DCMA Arms: | | | | | | | | | |
| 4 | 15 | 19 | 22 | 28 | 34 | 42 | 51 | 60 | 75 |
| 6 | 19 | 23 | 28 | 34 | 41 | 50 | 62 | 76 | 94 |
| 8 | 26 | 32 | 39 | 46 | 56 | 67 | 80 | 96 | 114 |
| 10 | 34 | 42 | 48 | 56 | 64 | 75 | 87 | 102 | 120 |
| 12 | 43 | 53 | 60 | 68 | 93 | 105 | 122 | 141 | 163 |
| 14 | 48 | 59 | 66 | 73 | 97 | 107 | 125 | 145 | 167 |
| 16 | 56 | 68 | 74 | 80 | 103 | 111 | 130 | 153 | 180 |
| 18 | 61 | 75 | 80 | 84 | 107 | 114 | 133 | 156 | 184 |
| 20 | 70 | 87 | 92 | 97 | 122 | 129 | 139 | 163 | 191 |
| 24 | 84 | 103 | 106 | 108 | 132 | 140 | 150 | 174 | 201 |
| PCMA Arms: | | | | | | | | | |
| 6 | 25 | 31 | 38 | 46 | 56 | — | — | — | — |
| 8 | 35 | 44 | 53 | 65 | 80 | — | — | — | — |
| 10 | 46 | 57 | 69 | 85 | 103 | — | — | — | — |
| 12 | 58 | 71 | 87 | 107 | 130 | — | — | — | — |
| HMLA Arms: | | | | | | | | | |
| 6 | 34 | 41 | 50 | 62 | 75 | — | — | — | — |
| 8 | 47 | 58 | 71 | 87 | 106 | — | — | — | — |
| 10 | 62 | 75 | 92 | 113 | 139 | — | — | — | — |
| 12 | 78 | 96 | 118 | 144 | 176 | — | — | — | — |
| CMA Arms: | | | | | | | | | |
| 4 | 13 | 16 | 20 | 25 | 30 | 39 | 42 | 43 | 49 |
| 6 | 17 | 21 | 26 | 31 | 38 | 41 | 46 | 51 | 58 |
| 8 | 24 | 30 | 36 | 45 | 55 | 58 | 63 | 70 | 79 |
| 10 | 31 | 38 | 46 | 57 | 69 | 73 | 78 | 86 | 95 |
| 12 | 39 | 48 | 59 | 72 | 88 | 92 | 98 | 105 | 115 |
| 14 | 43 | 53 | 64 | 76 | 89 | 94 | 99 | 108 | 118 |
| 16 | 50 | 62 | 76 | 90 | 94 | 97 | 103 | 112 | 121 |
| 18 | 55 | 68 | 82 | 98 | 101 | 104 | 111 | 118 | 126 |
| 20 | 63 | 77 | 94 | 112 | 114 | 116 | 124 | 130 | 138 |
| 24 | 76 | 93 | 113 | 114 | 116 | 120 | 126 | 134 | 139 |

The above hours are for the erection of arms. See page 232 for assembling of arm time requirement; for general notes and description of arms, see page 231.

# FIRE PROTECTION AND LIFE-SAVING EQUIPMENT

HOURS REQUIRED EACH

| Item Description | Hours Required |
|---|---|
| Fire Protection: | |
| Hose Reel and Hose (1½" × 100' Hose) | 2.0 |
| Hose Reel and Hose (1½" × 75' Hose) | 2.0 |
| Fire Monitors | 3.0 |
| $CO_2$ Bottles—15# and 30# | 1.0 |
| 27# Portable Dry Chemical Extinguisher | 1.0 |
| 150# Dry Chemical Extinguisher | 1.0 |
| 2000# Dry Chemical System | 6.0 |
| Life-Saving Equipment: | |
| Life Raft—Eight-Man | 1.0 |
| Life Rings | 0.5 |
| 23-Man Survival Capsule | 24.0 |
| Stretcher | 1.0 |
| First Aid Kit | 0.5 |

Above hours include break-loose of tie-down or checking out of barge storage as may be required, rigging, picking or handling and setting in position.

These hours are average for handling and installing the items listed and may need adjusting depending on conditions and circumstances.

# ZINC BRACELET ANODES

HOURS REQUIRED EACH

| Pipe Size O.D. | Thickness of Coating | Net Weight Pounds | Gross Weight Pounds | Hours Required Each |
|---|---|---|---|---|
| CYLINDRICAL TYPE | | | | |
| 10¾″ | ⅛″ | 160 | 165 | 3.00 |
| 10¾″ | ⅛″ | 160 | 171 | 3.00 |
| 10¾″ | 5/32″ | 260 | 273 | 3.00 |
| 12¾″ | ½″ | 240 | 235 | 3.25 |
| 14″ | 5/32″ | 240 | 254 | 3.50 |
| 16″ | 9/16″ | 290 | 306 | 3.75 |
| 16″ | 5/32″ | 360 | 376 | 3.75 |
| 18″ | 5/32″ | 470 | 488 | 4.00 |
| 20″ | ⅝″ | 375 | 394 | 4.25 |
| 20″ | 5/32″ | 375 | 394 | 4.25 |
| 24″ | ⅝″ | 450 | 473 | 4.50 |
| 24″ | 5/32″ | 450 | 473 | 4.50 |
| 26″ | 5/32″ | 525 | 550 | 4.75 |
| 36″ | ⅝″ | 850 | 900 | 5.00 |
| TAPERED CYLINDRICAL TYPE | | | | |
| 3½″ | 0.040″ | 36 | 38 | 2.00 |
| 4½″ | 0.040″ | 56 | 58 | 2.00 |
| 6⅝″ | 0.040″ | 80 | 84 | 2.00 |
| 8⅝″ | 0.040″ | 96 | 101 | 2.00 |
| 10¾″ | 0.125″ | 125 | 132 | 2.50 |
| 12¾″ | 0.500″ | 220 | 233 | 2.75 |

Above manhours are for rigging, picking, installing and securing in place the type and size anodes listed.

Net weight refers to total zinc alloy weight.

Gross weight refers to total zinc alloy and steel core weight.

# INSTALLATION OF GALVALUM ANODES

HOURS REQUIRED EACH

| Item No. | Galvalum Anode Description | Gross Weight Pounds | Hours Required Each |
|---|---|---|---|
| 1 | 325 lb.—8′ Anode with 2″ Straight Pipe Core ... | 375 | 2.50 |
| 2 | 325 lb.—8′ Anode with Steel Strap Core and 2″ Pipe Standoffs ..... | 357 | 2.50 |
| 3 | 325 lb.—8′ Anode with 2″ Offset Pipe Core ...... | 385 | 3.00 |
| 4 | 325 lb.—8′ Anode with 2″ Offset Pipe Core and Gussets ..... | — | 6.00 |
| 5 | 725 lb.—8′ Anode with 4″ Offset Pipe Core ...... | 909 | 4.00 |
| 6 | 325 lb.—Dual 8′ Anodes, 2″ Offset Pipe Cores and Clamps ..... | — | 3.50 |
| 7 | 325 lb.—8′ Anode with 2″ Pipe Core and Standoffs and Gussets ..... | — | 6.00 |
| 8 | 325 lb.—8′ Anode with Angle Iron Core and 2″ Pipe Standoffs and 6″ Channel ..... | 367 | 6.50 |
| 9 | 325 lb.—8′ Anode with 2″ Offset Pipe Core and Bolt-On Mounting Assembly ..... | 385 | 3.00 |
| 10 | 325 lb.—Up to 8′ Long Anode on 1″ Stud, 2″ Pipe Spacer ..... | — | 5.00 |
| 11 | 325 lb.—Up to 8′ Long Anode Flush-Mounted with 1″ Stub Bolt ..... | — | 4.75 |

Above hours include rigging, picking, setting, aligning and bolting or welding in place as required.

Weights shown include aluminum alloy and steel weight.

# HANDLING AND ERECTING PREFABRICATED SPOOLED PIPING

HOURS REQUIRED PER LINEAR FOOT BY SIZE

| Pipe Size Inches | PIPE SCHEDULE NUMBERS | | |
|---|---|---|---|
| | 10 to 60 | 80 to 100 | 120 to 160 |
| ½ | 0.26 | 0.29 | 0.30 |
| ⅜ | 0.27 | 0.29 | 0.32 |
| ½ | 0.27 | 0.30 | 0.34 |
| ¾ | 0.28 | 0.32 | 0.35 |
| 1 | 0.29 | 0.34 | 0.39 |
| 1¼ | 0.30 | 0.35 | 0.41 |
| 1½ | 0.32 | 0.37 | 0.45 |
| 2 | 0.34 | 0.40 | 0.49 |
| 2½ | 0.36 | 0.44 | 0.54 |
| 3 | 0.39 | 0.48 | 0.59 |
| 3½ | 0.40 | 0.50 | 0.62 |
| 4 | 0.41 | 0.52 | 0.66 |
| 5 | 0.44 | 0.57 | 0.72 |
| 6 | 0.47 | 0.64 | 0.84 |
| 8 | 0.57 | 0.81 | 0.99 |
| 10 | 0.72 | 1.00 | 1.38 |
| 12 | 0.88 | 1.23 | 1.69 |
| 14 O.D. | 1.01 | 1.46 | 2.01 |
| 16 O.D. | 1.27 | 1.71 | 2.34 |
| 18 O.D. | 1.48 | 1.96 | 2.69 |
| 20 O.D. | 1.74 | 2.22 | 3.04 |
| 24 O.D. | 1.94 | 2.51 | 3.43 |

Above hours include rigging, picking, setting in place, and aligning of carbon steel pipe spools.

The above time frames do not include welding bolt-up or screwed make-up. See other tables for these required hours.

Units apply to any length spool piece or segment of work.

# MAKING UP SCREWED FITTINGS AND VALVES

HOURS REQUIRED FOR EACH MAKE-UP

| Nominal Pipe Size—Inches | ALL SCHEDULES—PER CONNECTION | |
|---|---|---|
| | Plain | Back Welded |
| ¼ | 0.1 | 0.4 |
| ⅜ | 0.1 | 0.4 |
| ½ | 0.1 | 0.4 |
| ¾ | 0.1 | 0.5 |
| 1 | 0.2 | 0.5 |
| 1¼ | 0.2 | 0.6 |
| 1½ | 0.3 | 0.7 |
| 2 | 0.3 | 0.9 |
| 2½ | 0.4 | 1.0 |
| 3 | 0.4 | 1.2 |
| 3½ | 0.4 | 1.4 |
| 4 | 0.5 | 1.6 |

Above time frames are for individual connection only. For cutting, threading, handling and erecting, additional hours are required. See other tables pertaining to these operations.

Ells and Valves = Two Connections
Tees = Three Connections
Crosses = Four Connections

# HANDLE AND POSITION VALVES

HOURS REQUIRED EACH

| Pipe Size Inches | SERVICE PRESSURE RATING | | | | |
|---|---|---|---|---|---|
| | 150 lb. | 300-400 lb. | 600-900 lb. | 1500 lb. | 2500 lb. |
| ¼ | 0.2 | 0.2 | 0.2 | 0.4 | 0.4 |
| ⅜ | 0.2 | 0.2 | 0.3 | 0.4 | 0.5 |
| ½ | 0.2 | 0.2 | 0.3 | 0.5 | 0.5 |
| ¾ | 0.2 | 0.3 | 0.5 | 0.6 | 0.6 |
| 1 | 0.3 | 0.3 | 0.6 | 0.6 | 0.7 |
| 1¼ | 0.3 | 0.3 | 0.7 | 0.9 | 1.2 |
| 1½ | 0.4 | 0.4 | 1.0 | 1.2 | 1.4 |
| 2 | 0.5 | 0.8 | 1.3 | 1.5 | 1.8 |
| 2½ | 0.8 | 1.1 | 1.5 | 1.9 | 2.1 |
| 3 | 1.2 | 1.5 | 2.0 | 2.4 | 2.6 |
| 3½ | 1.4 | 1.7 | 2.3 | 2.7 | 2.9 |
| 4 | 1.7 | 2.0 | 2.6 | 3.1 | 3.4 |
| 5 | 2.0 | 2.4 | 3.0 | 3.6 | 4.0 |
| 6 | 2.2 | 2.7 | 3.3 | 4.1 | 4.1 |
| 8 | 2.8 | 3.4 | 4.2 | 5.3 | 5.9 |
| 10 | 3.6 | 4.2 | 5.1 | 6.8 | 7.0 |
| 12 | 4.3 | 5.1 | 6.3 | 8.5 | 9.5 |
| 14 | 5.1 | 6.0 | 7.5 | 10.5 | 11.4 |
| 16 | 5.9 | 7.1 | 8.8 | 12.7 | 13.0 |
| 18 | 6.7 | 8.1 | 10.4 | 15.1 | 15.8 |
| 20 | 7.7 | 9.2 | 11.9 | 17.9 | 18.4 |
| 24 | 8.5 | 10.3 | 13.6 | 20.7 | 21.5 |

Above hours include rigging, picking setting in place, and aligning of valves and expansion joints.

Time frames do not include welding, bolting-up or making-up of valve joints. See other tables for these charges.

Use 150-lb. allowance for standard brass and iron valves.

Use 300-lb. allowance for extra heavy and 200-lb. brass and iron valves.

For motor operated or diaphragm valves, add 125 percent to the above hours.

# FLANGED BOLT-UPS

HOURS REQUIRED EACH

| Pipe Size Inches | SERVICE PRESSURE RATING | | | | | |
|---|---|---|---|---|---|---|
| | 150 lb. | 300-400 lb. | 600 lb. | 900 lb. | 1500 lb. | 2500 lb. |
| 2 or Less | 0.7 | 0.8 | 0.9 | 1.0 | 1.2 | 1.6 |
| 2½ | 0.8 | 0.9 | 1.0 | 1.2 | 1.5 | 2.0 |
| 3 | 0.8 | 0.9 | 1.0 | 1.2 | 1.5 | 2.0 |
| 3½ | 1.0 | 1.2 | 1.3 | 1.5 | 1.8 | 2.4 |
| 4 | 1.2 | 1.4 | 1.5 | 1.7 | 2.1 | 2.8 |
| 6 | 1.5 | 1.7 | 1.8 | 2.1 | 2.6 | 3.4 |
| 8 | 2.1 | 2.4 | 2.6 | 3.0 | 3.7 | 4.9 |
| 10 | 2.7 | 3.0 | 3.2 | 3.7 | 4.6 | 6.1 |
| 12 | 3.4 | 3.8 | 4.1 | 4.7 | 5.8 | 7.7 |
| 14 | 3.8 | 4.3 | 4.6 | 5.3 | 6.5 | 8.5 |
| 16 | 4.4 | 4.9 | 5.2 | 6.0 | 7.4 | 10.0 |
| 18 | 4.8 | 5.4 | 5.8 | 6.7 | 8.2 | 11.5 |
| 20 | 5.5 | 6.2 | 6.6 | 7.6 | 9.3 | 12.8 |
| 24 | 6.6 | 7.4 | 7.9 | 9.1 | 11.2 | 14.5 |

Above hours are for obtaining from storage and handling of bolts or studs and gaskets and bolting up of each joint.

Time frames do not include handling of valves, flanged fittings or spools. See other tables for these hours.

Where tongue and groove, ring joint, female or fittings with special facings are used, add 25 percent to the above hours.

For standard cast iron use 150-lb. allowance.

For extra heavy cast iron use 300-lb. allowance.

# GENERAL WELDING NOTES

*Backing Rings:* When backing rings are used, add 25% to the welding man hours to cover extra problems in fit-up. In addition the following percentages should be added if applicable:

1) When backing rings are tack welded in on one side, add 10% to the man hours of a standard thickness butt weld.
2) When backing rings are completely welded in on one side, add 30% to the man hours of a standard thickness butt weld.
3) Preheating and stress relieving, when required, should be charged at full butt weld preheating and stress relieving man hours for the size and thickness in which the backing ring is installed.

*Nozzle Welds:* Following percentage increases should be allowed for the following conditions:

1) When nozzle welds are to be located off-center of the run(except tangential) increase man hours shown for nozzle welds, 50%.
2) Add 80% to nozzle welds for tangential nozzle welds.
3) When nozzle welds are to be located on a fitting increase nozzle weld man hours 50%.

*Long Neck Nozzle Welds:* The welding-on of long neck nozzles should be charged at the schedule 160 reinforced nozzle weld man hours.

*Shaped Nozzles, Nozzle Weld Fit-Ups and Dummy Nozzle Welds:* These should be charged at a percentage of the completed nozzle weld man hours as follows:

1) Shaped Branch ........ 50%
2) Shaped Hole in Header ........ 50%
3) Fit-up of Both Branch or Header (whether tack-welded or not) ........ 60%
4) Dummy Nozzle Weld (no holes in header) ........ 70%

*Sloping Lines:* Add 100% to all welding man hours for this condition.

*Consumable Inserts:* When consumable inserts are used, add the following percentages to the welding man hours to cover extra problems in fit-up:

1) Through 1/2″ wall ........ 40%
2) Over 1/2″ through 1″ wall ........ 30%
3) Over 1″ through 2″ wall ........ 20%
4) Over 2″ through 3″ wall ........ 15%
5) Over 3″ wall ........ 10%

# MANUAL CARBON STEEL PIPE BUTT WELDS

HOURS REQUIRED FOR SIZES LISTED

| Pipe Size Inches | Standard Pipe & O.D. Sizes ⅜" Thick | Extra Heavy Pipe & O.D. Sizes ½" Thick | PIPE SCHEDULE NUMBERS | | | | | | | | |
|---|---|---|---|---|---|---|---|---|---|---|---|
| | | | 20 | 30 | 40 | 60 | 80 | 100 | 120 | 140 | 160 |
| 1 | 0.7 | 0.8 | — | — | 0.7 | — | 0.8 | — | — | — | 1.0 |
| 1¼ | 0.8 | 0.8 | — | — | 0.8 | — | 0.8 | — | — | — | 1.1 |
| 1½ | 0.8 | 0.9 | — | — | 0.8 | — | 0.9 | — | — | — | 1.3 |
| 2 | 1.0 | 1.0 | — | — | 1.0 | — | 1.0 | — | — | — | 1.6 |
| 2½ | 1.2 | 1.3 | — | — | 1.2 | — | 1.3 | — | — | — | 1.8 |
| 3 | 1.3 | 1.4 | — | — | 1.3 | — | 1.4 | — | — | — | 2.1 |
| 3½ | 1.4 | 1.6 | — | — | 1.4 | — | 1.6 | — | — | — | — |
| 4 | 1.5 | 1.8 | — | — | 1.5 | — | 1.8 | — | 2.8 | — | 3.0 |
| 5 | 1.7 | 2.1 | — | — | 1.7 | — | 2.1 | — | 2.9 | — | 3.8 |
| 6 | 2.0 | 2.5 | — | — | 2.0 | — | 2.5 | — | 3.8 | — | 4.9 |
| 8 | 2.6 | 3.3 | 2.6 | 2.6 | 2.6 | 3.0 | 3.3 | 4.6 | 6.0 | 7.5 | 8.6 |
| 10 | 3.1 | 4.0 | 3.1 | 3.1 | 3.1 | 4.0 | 5.1 | 6.8 | 9.4 | 11.4 | 13.1 |
| 12 | 3.6 | 4.7 | 3.6 | 3.6 | 4.1 | 5.2 | 6.6 | 9.9 | 12.2 | 15.3 | 17.9 |
| 14 O.D. | 4.3 | 5.7 | 4.3 | 4.3 | 5.0 | 6.8 | 9.6 | 13.2 | 16.2 | 19.2 | 22.7 |
| 16 O.D. | 5.0 | 6.6 | 5.0 | 5.0 | 6.6 | 8.4 | 12.4 | 19.5 | 20.7 | 25.0 | 27.7 |
| 18 O.D. | 5.9 | 7.7 | 5.9 | 6.8 | 8.6 | 11.2 | 16.4 | 21.8 | 25.6 | 29.9 | 33.7 |
| 20 O.D. | 6.3 | 8.4 | 6.3 | 8.4 | 9.4 | 13.8 | 19.5 | 26.0 | 31.9 | 37.0 | 40.8 |
| 24 O.D. | 6.9 | 10.1 | 6.9 | — | 13.3 | 20.1 | 25.2 | 35.8 | 43.5 | 49.3 | 59.3 |

The wall thickness of the pipe will determine the hours to be applied. For butt welds of double extra strong materials, use schedule 160 hours.

For mitre welds, add 50 percent to the above hours.

Above hours do not include cutting and beveling of pipe. See other tables for these time frames.

If preheating is specified or required by code, add for this operation. See respective table for these hours.

Stress relieving of welds in carbon steel materials is required by the A.S.A. code for pressure piping, where the wall thickness is ¾ inch or greater. All butt weld sizes shown below the ruled lines are ¾ inch or greater.

Where stress relieving is required, an extra charge should be made. See respective table for time required for this operation.

For additional notes on welding, see preceding page.

# 90° CARBON STEEL WELDED PIPE NOZZLES

HOURS REQUIRED FOR SIZES LISTED

| Pipe Size Inches | Standard Pipe & O.D. Sizes ⅜" Thick | Extra Heavy Pipe & O.D. Sizes ½" Thick | PIPE SCHEDULE NUMBERS | | | | | | | | |
|---|---|---|---|---|---|---|---|---|---|---|---|
| | | | 20 | 30 | 40 | 60 | 80 | 100 | 120 | 140 | 160 |
| 1 | 2.1 | 2.2 | — | — | 2.1 | — | 2.2 | — | — | — | 3.1 |
| 1¼ | 2.2 | 2.4 | — | — | 2.2 | — | 2.4 | — | — | — | 3.6 |
| 1½ | 2.4 | 2.6 | — | — | 2.4 | — | 2.6 | — | — | — | 4.0 |
| 2 | 2.5 | 3.1 | — | — | 2.5 | — | 3.1 | — | — | — | 5.3 |
| 2½ | 2.8 | 3.8 | — | — | 2.8 | — | 3.8 | — | — | — | 5.9 |
| 3 | 3.2 | 4.4 | — | — | 3.2 | — | 4.4 | — | — | — | 6.6 |
| 3½ | 3.7 | 4.9 | — | — | 3.7 | — | 4.9 | — | — | — | — |
| 4 | 4.0 | 5.6 | — | — | 4.0 | — | 5.6 | — | 7.0 | — | 8.6 |
| 5 | 5.1 | 6.9 | — | — | 5.1 | — | 6.9 | — | 8.6 | — | 10.7 |
| 6 | 5.4 | 7.5 | — | — | 5.4 | — | 7.5 | — | 10.9 | — | 13.9 |
| 8 | 6.3 | 8.9 | 6.3 | 6.3 | 6.3 | 8.3 | 8.9 | 12.0 | 15.2 | 18.6 | 21.5 |
| 10 | 7.1 | 10.3 | 7.1 | 7.1 | 7.1 | 10.3 | 12.6 | 16.4 | 21.1 | 27.3 | 32.8 |
| 12 | 8.1 | 11.8 | 8.1 | 8.1 | 9.9 | 13.1 | 17.0 | 23.5 | 28.7 | 34.7 | 39.2 |
| 14 O.D. | 9.3 | 13.6 | 9.3 | 9.3 | 11.6 | 16.0 | 22.7 | 28.9 | 34.6 | 38.9 | 47.9 |
| 16 O.D. | 10.6 | 15.2 | 10.6 | 10.6 | 15.2 | 20.2 | 26.8 | 36.5 | 41.4 | 45.7 | 55.3 |
| 18 O.D. | 11.6 | 16.3 | 11.6 | 15.1 | 19.1 | 25.4 | 30.1 | 44.1 | 49.0 | 53.0 | 69.2 |
| 20 O.D. | 13.0 | 18.3 | 13.0 | 18.3 | 22.3 | 32.6 | 35.4 | 51.0 | 55.9 | 60.8 | 77.7 |
| 24 O.D. | 14.2 | 19.8 | 14.2 | 21.2 | 27.7 | 41.7 | 46.0 | 64.8 | 69.7 | 77.8 | 90.9 |

All nozzles other than 90° should be charged at the hours shown for 45° nozzles.

Wall thickness of the pipe used for the nozzle determines the hours that will apply. For nozzles of double extra strong pipe thickness, use schedule 160 hours.

Hours given above are for plain welded nozzles only. For use of gussett plates, etc., as stiffeners, not for reinforcement, add 25 percent to the hours shown above.

If preheating is required, add for this operation. See respective table for these time frames. The size and wall thickness of the header (not the size of the nozzle) determines the preheating hours that apply.

If stress relieving is required, add for this operation. See respective table for required hours. Pipe sizes shown below the ruled line are ¾ inch or greater in wall thickness and should be stress relieved.

For additional notes on welding see page 241.

# 45° CARBON STEEL WELDED PIPE NOZZLES

HOURS REQUIRED FOR SIZES LISTED

| Pipe Size Inches | Standard Pipe & O.D. Sizes ⅜" Thick | Extra Heavy Pipe & O.D. Sizes ½" Thick | PIPE SCHEDULE NUMBERS | | | | | | | | |
|---|---|---|---|---|---|---|---|---|---|---|---|
| | | | 20 | 30 | 40 | 60 | 80 | 100 | 120 | 140 | 160 |
| 1 | 2.8 | 2.9 | — | — | 2.8 | — | 2.9 | — | — | — | 4.1 |
| 1¼ | 2.9 | 3.2 | — | — | 2.9 | — | 3.2 | — | — | — | 4.7 |
| 1½ | 3.2 | 3.6 | — | — | 3.2 | — | 3.6 | — | — | — | 5.4 |
| 2 | 3.3 | 4.1 | — | — | 3.3 | — | 4.1 | — | — | — | 7.1 |
| 2½ | 3.8 | 5.1 | — | — | 3.8 | — | 5.1 | — | — | — | 7.9 |
| 3 | 4.4 | 5.9 | — | — | 4.4 | — | 5.9 | — | — | — | 8.9 |
| 3½ | 4.9 | 6.4 | — | — | 4.9 | — | 6.4 | — | — | — | — |
| 4 | 5.5 | 7.6 | — | — | 5.5 | — | 7.6 | — | 9.4 | — | 11.4 |
| 5 | 6.7 | 9.1 | — | — | 6.7 | — | 9.1 | — | 11.7 | — | 14.4 |
| 6 | 7.1 | 10.0 | — | — | 7.1 | — | 10.0 | — | 14.5 | — | 18.4 |
| 8 | 8.6 | 11.7 | 8.6 | 8.6 | 8.6 | 11.1 | 11.7 | 15.9 | 20.1 | 25.1 | 28.6 |
| 10 | 9.6 | 13.7 | 9.6 | 9.6 | 9.6 | 13.7 | 16.6 | 21.8 | 28.1 | 36.2 | 43.7 |
| 12 | 10.9 | 15.7 | 10.9 | 10.9 | 13.0 | 17.5 | 22.9 | 24.2 | 38.6 | 45.7 | 52.2 |
| 14 O.D. | 12.4 | 18.2 | 12.4 | 12.4 | 15.3 | 21.4 | 30.1 | 38.4 | 45.9 | 51.6 | 64.2 |
| 16 O.D. | 14.2 | 20.2 | 14.2 | 14.2 | 20.2 | 26.8 | 35.3 | 48.7 | 54.8 | 61.2 | 74.1 |
| 18 O.D. | 15.5 | 21.6 | 15.5 | 20.1 | 25.5 | 34.3 | 40.6 | 59.1 | 65.7 | 71.0 | 92.7 |
| 20 O.D. | 17.1 | 23.7 | 17.1 | 22.7 | 29.6 | 43.3 | 46.9 | 68.3 | 74.9 | 81.5 | 104.1 |
| 24 O.D. | 18.8 | 26.6 | 18.8 | 31.8 | 35.5 | 55.2 | 55.0 | 86.8 | 93.4 | 104.3 | 121.8 |

Wall thickness of the pipe used for the nozzle determines the hours that will apply. For nozzles of double extra strong pipe thickness, use schedule 160 hours.

Hours given above are for plain welder nozzles only. For use of gussett plates, etc., as stiffeners, not for reinforcement, add 25 percent to the hours shown above.

If preheating is required, add for this operation. See respective table for these time frames. The size and wall thickness of the header (not the size of the nozzle) determines the preheating hours that apply.

If stress relieving is required, add for this operation. See respective table for required hours. Pipe sizes shown below the ruled line are ¾ inch or greater in wall thickness and should be stress relieved.

For additional notes on welding see page 241.

# 90° COUPLING WELDS AND SOCKET WELDS

HOURS REQUIRED EACH FOR SIZES LISTED

| Pipe Size Inches | 90°-3000 lb. Coupling Weld | 90°-6000 lb. Coupling Weld | SOCKET WELDS | |
|---|---|---|---|---|
| | | | Sch. 40 & 80 Pipe | Sch. 100 & Heavier Pipe |
| ½" or Less | 1.6 | 2.0 | 0.6 | 0.6 |
| ¾ | 1.8 | 2.2 | 0.6 | 0.7 |
| 1 | 2.1 | 2.5 | 0.7 | 0.8 |
| 1¼ | 2.4 | 2.9 | 0.9 | 1.0 |
| 1½ | 2.6 | 3.2 | 0.9 | 1.2 |
| 2 | 3.3 | 4.1 | 1.0 | 1.5 |
| 2½ | 3.9 | 4.8 | 1.3 | 1.6 |
| 3 | 4.6 | 5.6 | 1.4 | 2.0 |

Hours shown are for welding of coupling to the O.D. of the pipe only.

If couplings are to be welded to the I.D. of the pipe, add 50 percent to the above hours for pipe thickness up to 1 inch, and an additional 12 percent for each ¼ inch or fraction thereof of pipe thickness over 1 inch.

Any coupling welded to pipe heavier than schedule 160 should be houred as a 6000-lb. coupling.

For couplings welded at angles from 45° to less than 90° and couplings attached to fittings, increase above hours 50 percent.

For couplings welded at angles less than 45°, increase above hours 75 percent.

Socket welds do not include cut. See respective table for these hours.

# PLAIN END FLAME CUTTING PIPE

HOURS REQUIRED FOR SIZES LISTED

| Pipe Size Inches | Standard Pipe & O.D. Sizes ⅜" Thick | Extra Heavy Pipe & O.D. Sizes ½" Thick | PIPE SCHEDULE NUMBERS | | | | | | | | |
|---|---|---|---|---|---|---|---|---|---|---|---|
| | | | 20 | 30 | 40 | 60 | 80 | 100 | 120 | 140 | 160 |
| 2" or Less | 0.10 | 0.15 | — | — | 0.10 | — | 0.15 | — | — | — | 0.21 |
| 2½ | 0.12 | 0.17 | — | — | 0.12 | — | 0.17 | — | — | — | 0.23 |
| 3 | 0.15 | 0.21 | — | — | 0.15 | — | 0.21 | — | — | — | 0.28 |
| 4 | 0.21 | 0.28 | — | — | 0.21 | — | 0.28 | — | 0.38 | — | 0.41 |
| 5 | 0.24 | 0.36 | — | — | 0.24 | — | 0.36 | — | 0.44 | — | 0.49 |
| 6 | 0.33 | 0.45 | — | — | 0.33 | — | 0.45 | — | 0.56 | — | 0.63 |
| 8 | 0.46 | 0.64 | 0.46 | 0.46 | 0.46 | 0.59 | 0.64 | 0.76 | 0.86 | 0.97 | 1.14 |
| 10 | 0.64 | 0.92 | 0.64 | 0.64 | 0.64 | 0.92 | 0.99 | 1.09 | 1.24 | 1.43 | 1.73 |
| 12 | 0.70 | 1.09 | 0.70 | 0.70 | 0.86 | 1.30 | 1.37 | 1.48 | 1.73 | 1.91 | 2.05 |
| 14 O.D. | 0.98 | 1.30 | 0.98 | 0.98 | 1.15 | 1.44 | 1.67 | 1.78 | 1.96 | 2.30 | 2.42 |
| 16 O.D. | 1.09 | 1.61 | 1.09 | 1.09 | 1.61 | 1.78 | 1.90 | 2.13 | 2.30 | 2.59 | 2.93 |
| 18 O.D. | 1.42 | 2.01 | 1.42 | 1.65 | 2.01 | 2.24 | 2.36 | 2.66 | 2.83 | 3.19 | 3.72 |
| 20 O.D. | 1.71 | 2.24 | 1.71 | 2.30 | 2.48 | 2.66 | 2.83 | 3.13 | 3.30 | 3.84 | 4.37 |
| 24 O.D. | 2.60 | 3.30 | 2.60 | 3.48 | 3.66 | 3.84 | 3.95 | 4.31 | 4.78 | 5.37 | 6.08 |

Above hours are for flame cutting carbon steel pipe materials.

For mitre cuts less than 30°, add 50 percent to the above hours.

For mitre cuts 30° or greater, add 100 percent to the above hours.

Hours are for cutting pipe with plain ends only and do not include beveling, threading, etc. See respective tables for these operations and time frames.

For cutting the ends of bends or trimming fittings, add 50 percent to the above hours.

# FLAME BEVELING PIPE FOR WELDING

HOURS REQUIRED FOR SIZES LISTED

| Pipe Size Inches | Standard Pipe & O.D. Sizes ⅜" Thick | Extra Heavy Pipe & O.D. Sizes ½" Thick | PIPE SCHEDULE NUMBERS | | | | | | | | |
|---|---|---|---|---|---|---|---|---|---|---|---|
| | | | 20 | 30 | 40 | 60 | 80 | 100 | 120 | 140 | 160 |
| 2" or Less | 0.08 | 0.12 | — | — | 0.08 | — | 0.12 | — | — | — | 0.16 |
| 2½ | 0.09 | 0.14 | — | — | 0.09 | — | 0.14 | — | — | — | 0.18 |
| 3 | 0.12 | 0.16 | — | — | 0.12 | — | 0.16 | — | — | — | 0.22 |
| 4 | 0.16 | 0.22 | — | — | 0.16 | — | 0.22 | — | 0.30 | — | 0.32 |
| 5 | 0.20 | 0.28 | — | — | 0.20 | — | 0.28 | — | 0.35 | — | 0.39 |
| 6 | 0.26 | 0.35 | — | — | 0.26 | — | 0.35 | — | 0.45 | — | 0.49 |
| 8 | 0.37 | 0.51 | 0.37 | 0.37 | 0.37 | 0.46 | 0.51 | 0.60 | 0.68 | 0.75 | — |
| 10 | 0.51 | 0.72 | 0.51 | 0.51 | 0.51 | 0.72 | 0.78 | 0.86 | 0.95 | — | — |
| 12 | 0.55 | 0.86 | 0.55 | 0.55 | 0.68 | 1.02 | 1.08 | 1.18 | — | — | — |
| 14 O.D. | 0.57 | 1.02 | 0.77 | 0.77 | 0.91 | 1.13 | 1.31 | — | — | — | — |
| 16 O.D. | 0.86 | 1.27 | 0.86 | 0.86 | 1.27 | 1.40 | 1.55 | — | — | — | — |
| 18 O.D. | 1.11 | 1.58 | 1.11 | 1.30 | 1.58 | 1.77 | — | — | — | — | — |
| 20 O.D. | 1.35 | 1.77 | 1.35 | 1.82 | 1.95 | 2.15 | — | — | — | — | — |
| 24 O.D. | 2.04 | 2.60 | 2.04 | 2.74 | 2.88 | — | — | — | — | — | — |

Above hours are for beveling carbon steel pipe materials.

For mitre bevels add 50 percent to the above hours.

Above hours are for flame "V" beveling only and do not include cutting. See respective table for these time requirements.

For beveling on the ends of bends or trimmed fittings, add 50 percent to the above hours.

# CUTTING AND THREADING PIPE

HOURS REQUIRED FOR SIZES LISTED

| Pipe Size Inches | Standard Pipe & O.D. Sizes ⅜" Thick | Extra Heavy Pipe & O.D. Sizes ½" Thick | PIPE SCHEDULE NUMBERS | | | | | | | | |
|---|---|---|---|---|---|---|---|---|---|---|---|
| | | | 20 | 30 | 40 | 60 | 80 | 100 | 120 | 140 | 160 |
| 2" or Less | 0.20 | 0.30 | — | — | 0.20 | — | 0.30 | — | — | — | 0.42 |
| 2½ | 0.27 | 0.40 | — | — | 0.27 | — | 0.40 | — | — | — | 0.47 |
| 3 | 0.30 | 0.42 | — | — | 0.30 | — | 0.42 | — | — | — | 0.58 |
| 4 | 0.42 | 0.58 | — | — | 0.42 | — | 0.58 | — | 0.84 | — | 0.90 |
| 5 | 0.54 | 0.78 | — | — | 0.54 | — | 0.78 | — | 0.91 | — | 1.07 |
| 6 | 0.70 | 0.93 | — | — | 0.70 | — | 0.93 | — | 1.18 | — | 1.35 |
| 8 | 0.97 | 1.35 | 0.97 | 0.97 | 0.97 | 1.26 | 1.35 | 1.64 | 1.83 | 2.04 | 2.38 |
| 10 | 1.38 | 1.84 | 1.38 | 1.38 | 1.38 | 1.84 | 2.11 | 2.44 | 2.58 | 3.08 | 3.65 |
| 12 | 1.53 | 2.44 | 1.53 | 1.53 | 2.07 | 2.75 | 2.90 | 3.17 | 3.65 | 4.14 | 4.38 |
| 14 O.D. | 2.07 | 2.67 | 2.07 | 2.07 | 2.67 | 2.95 | 3.39 | 3.85 | 4.14 | — | — |
| 16 O.D. | 2.37 | 3.45 | 2.37 | 2.37 | 3.45 | 3.71 | 4.14 | 4.38 | 4.87 | — | — |
| 18 O.D. | 2.95 | 4.14 | 2.95 | 3.65 | 4.14 | 4.38 | 4.87 | 5.42 | — | — | — |
| 20 O.D. | 3.54 | 4.72 | 3.54 | 4.72 | 5.68 | 6.16 | 6.49 | — | — | — | — |
| 24 O.D. | 5.18 | 6.89 | 5.18 | 7.27 | 7.67 | 7.93 | 8.37 | — | — | — | — |

Above hours are for cutting and threading carbon steel pipe materials and threading is for die cut IPS threads only.

For threading the ends of bends, add 100 percent to the above hours.

# PREHEATING BUTT AND FLANGE WELDS

HOURS REQUIRED FOR SIZES LISTED

| Pipe Size Inches | Standard Pipe & O.D. Sizes ⅜" Thick | Extra Heavy Pipe & O.D. Sizes ½" Thick | PIPE SCHEDULE NUMBERS | | | | | | | | |
|---|---|---|---|---|---|---|---|---|---|---|---|
| | | | 20 | 30 | 40 | 60 | 80 | 100 | 120 | 140 | 160 |
| 2 | 0.2 | 0.3 | — | — | 0.2 | — | 0.3 | — | — | — | 0.5 |
| 2½ | 0.3 | 0.5 | — | — | 0.3 | — | 0.5 | — | — | — | 0.6 |
| 3 | 0.5 | 0.6 | — | — | 0.5 | — | 0.6 | — | — | — | 0.7 |
| 3½ | 0.5 | 0.6 | — | — | 0.5 | — | 0.6 | — | — | — | 0.9 |
| 4 | 0.6 | 0.7 | — | — | 0.6 | — | 0.7 | — | 0.9 | — | 0.9 |
| 5 | 0.7 | 0.9 | — | — | 0.7 | — | 0.9 | — | 0.9 | — | 1.0 |
| 6 | 0.8 | 1.0 | — | — | 0.8 | — | 1.0 | — | 1.3 | — | 1.5 |
| 8 | 0.9 | 1.3 | 0.9 | 0.9 | 0.9 | 1.3 | 1.3 | 1.8 | 1.9 | 2.4 | 2.5 |
| 10 | 1.3 | 1.8 | 1.3 | 1.3 | 1.3 | 1.8 | 2.0 | 2.4 | 2.7 | 3.3 | 3.8 |
| 12 | 1.5 | 2.0 | 1.5 | 1.5 | 1.9 | 2.2 | 2.8 | 3.3 | 3.8 | 4.4 | 5.3 |
| 14 O.D. | 1.9 | 2.5 | 1.9 | 1.9 | 2.2 | 3.0 | 3.5 | 4.4 | 5.0 | 5.8 | 6.6 |
| 16 O.D. | 2.2 | 3.3 | 2.2 | 2.2 | 3.0 | 3.8 | 4.5 | 5.4 | 6.0 | 7.3 | 8.5 |
| 18 O.D. | 2.6 | 3.5 | 2.6 | 3.1 | 4.1 | 5.0 | 6.0 | 7.0 | 7.9 | 8.5 | 10.5 |
| 20 O.D. | 3.1 | 4.1 | 3.1 | 4.1 | 5.2 | 6.3 | 7.4 | 8.7 | 9.8 | 11.1 | 12.9 |
| 24 O.D. | 3.7 | 5.0 | 3.7 | 5.3 | 6.4 | 7.8 | 9.3 | 10.4 | 11.7 | 13.3 | 15.2 |

Above hours are for preheating butt and flange welds of carbon steel or alloy materials up to 400° F.

For preheating to temperatures above 400° F. but not exceeding 600° F., add 100 percent to the above hours.

For preheating mitre welds, add 50 percent to the above hours.

For preheating double extra strong material, use schedule 160 hours.

# PREHEATING NOZZLE WELDS

HOURS REQUIRED FOR SIZES LISTED

| Pipe Size Inches | Standard Pipe & O.D. Sizes ⅜" Thick | Extra Heavy Pipe & O.D. Sizes ½" Thick | PIPE SCHEDULE NUMBERS | | | | | | | | |
|---|---|---|---|---|---|---|---|---|---|---|---|
| | | | 20 | 30 | 40 | 60 | 80 | 100 | 120 | 140 | 160 |
| 2 | 0.5 | 0.6 | — | — | 0.5 | — | 0.6 | — | — | — | 0.7 |
| 2½ | 0.6 | 0.7 | — | — | 0.6 | — | 0.7 | — | — | — | 0.9 |
| 3 | 0.6 | 0.9 | — | — | 0.6 | — | 0.9 | — | — | — | 1.0 |
| 3½ | 0.7 | 0.9 | — | — | 0.7 | — | 0.9 | — | — | — | — |
| 4 | 0.9 | 1.0 | — | — | 0.9 | — | 1.0 | — | 1.4 | — | 1.7 |
| 5 | 1.0 | 1.4 | — | — | 1.0 | — | 1.4 | — | 1.7 | — | 1.8 |
| 6 | 1.3 | 1.8 | — | — | 1.3 | — | 1.8 | — | 2.0 | — | 2.4 |
| 8 | 1.8 | 2.1 | 1.7 | 1.7 | 1.7 | 2.0 | 2.1 | 2.7 | 3.1 | 3.5 | 4.1 |
| 10 | 2.0 | 2.7 | 2.0 | 2.0 | 2.0 | 2.7 | 3.1 | 3.8 | 4.5 | — | 5.9 |
| 12 | 2.5 | 3.3 | 2.5 | 2.5 | 2.8 | 3.5 | 4.5 | 5.2 | 6.0 | — | 8.3 |
| 14 O.D. | 3.0 | 3.8 | 3.0 | 3.0 | 3.5 | 4.6 | 5.9 | 6.6 | 7.8 | — | 10.5 |
| 16 O.D. | 3.4 | 4.6 | 3.4 | 3.4 | 4.6 | 5.9 | 7.2 | 8.5 | 9.8 | — | 13.7 |
| 18 O.D. | 4.2 | 5.5 | 4.2 | 4.8 | 6.3 | 7.9 | 9.6 | 10.5 | 12.7 | — | 17.0 |
| 20 O.D. | 5.0 | 6.5 | 5.0 | 6.5 | 8.3 | 10.1 | 11.8 | 13.7 | 15.6 | — | — |
| 24 O.D. | 6.0 | 7.8 | 6.0 | 8.5 | 10.3 | 12.5 | 15.1 | 16.5 | 18.9 | — | — |

Above hours are for preheating 90° nozzle welds of carbon steel or alloy materials up to 400° F.

For preheating to temperatures above 400° F. but not exceeding 600° F., add 100 percent to the above hours.

The size of the nozzle and the wall thickness of the header or nozzle (whichever is greater) determines the hours to be used. For preheating of double extra strong thickness, use schedule 160 hours.

For preheating 45° nozzle welds, add 50 percent to the above hours.

Preheating of coupling, weld-o-let, thread-o-let or socket welds should be charged at the same hours as shown for the same size and schedule nozzle.

# STRESS RELIEVING BUTT, FLANGE AND NOZZLE WELDS

HOURS REQUIRED FOR SIZES LISTED

| Pipe Size Inches | Standard Pipe & O.D. Sizes ⅜" Thick | Extra Heavy Pipe & O.D. Sizes ½" Thick | PIPE SCHEDULE NUMBERS | | | | | | | | |
|---|---|---|---|---|---|---|---|---|---|---|---|
| | | | 20 | 30 | 40 | 60 | 80 | 100 | 120 | 140 | 160 |
| 2 | 2.6 | 2.8 | — | — | 2.6 | — | 2.8 | — | — | — | 3.0 |
| 2½ | 2.8 | 2.9 | — | — | 2.8 | — | 2.9 | — | — | — | 3.1 |
| 3 | 2.9 | 3.0 | — | — | 2.9 | — | 3.0 | — | — | — | 3.5 |
| 3½ | 3.0 | 3.1 | — | — | 3.0 | — | 3.1 | — | — | — | 3.8 |
| 4 | 3.0 | 3.5 | — | — | 3.0 | — | 3.5 | — | 3.6 | — | 3.9 |
| 5 | 3.5 | 3.7 | — | — | 3.5 | — | 3.7 | — | 4.1 | — | 4.3 |
| 6 | 3.7 | 4.1 | — | — | 3.7 | — | 4.1 | — | 4.3 | — | 4.9 |
| 8 | 4.2 | 4.7 | 4.2 | 4.2 | 4.2 | 4.4 | 4.7 | 5.1 | 5.3 | 5.5 | 5.9 |
| 10 | 4.6 | 5.1 | 4.6 | 4.6 | 4.6 | 5.1 | 5.3 | 5.7 | 5.9 | 6.3 | 6.7 |
| 12 | 5.1 | 5.5 | 5.1 | 5.1 | 5.3 | 5.8 | 6.0 | 6.5 | 6.8 | 7.1 | 7.4 |
| 14 O.D. | 5.5 | 5.9 | 5.5 | 5.5 | 5.9 | 6.3 | 6.7 | 7.1 | 7.6 | 7.9 | 8.3 |
| 16 O.D. | 5.9 | 6.4 | 5.9 | 5.9 | 6.4 | 6.8 | 7.2 | 7.8 | 8.0 | 8.5 | 9.2 |
| 18 O.D. | 6.4 | 6.8 | 6.4 | 6.6 | 6.8 | 7.3 | 7.8 | 8.3 | 8.7 | 9.2 | 10.1 |
| 20 O.D. | 6.6 | 7.0 | 6.6 | 6.8 | 7.3 | 7.8 | 8.3 | 9.2 | 9.6 | 10.0 | 11.1 |
| 24 O.D. | 7.1 | 7.3 | 7.1 | 7.6 | 8.0 | 8.5 | 9.2 | 10.1 | 10.5 | 11.2 | 12.5 |

Above hours are for stress relieving butt, flange and nozzle welds in carbon steel materials.

For stress relieving butt welds and flange welds, the wall thickness of the pipe determines the hours that apply. For stress relieving nozzle welds, the size and thickness of the header to which the nozzle is attached determines the hours that apply. For stress relieving of double extra strong material, use schedule 160 hours.

All welds in piping materials having a wall thickness of ¾ inch or greater should be stress relieved to comply with the requirements of the A.S.A. code for pressure piping. Hours shown below the ruled line in the above schedule cover sizes having a wall thickness of ¾ inch or greater.

# RADIOGRAPHIC INSPECTION

## X-Ray Pipe Welds

| Nominal Pipe Size | Wall Thickness Through Extra Strong | Wall Thickness Greater Than Extra Strong Through Schedule 120 | Wall Thickness Greater Than Schedule 120 Through Double Extra Strong |
|---|---|---|---|
| 2″ or Less | 0.86 | — | 1.13 |
| 3 | 0.86 | — | 1.13 |
| 4 | 0.98 | 1.13 | 1.27 |
| 5 | 1.07 | 1.23 | 1.40 |
| 6 | 1.20 | 1.40 | 1.56 |
| 8 | 1.35 | 1.54 | 1.75 |
| 10 | 1.51 | 1.73 | 1.97 |
| 12 | 1.71 | 1.97 | 2.23 |
| 14 | 1.86 | 2.14 | 2.42 |
| 16 | 2.08 | 2.39 | 2.70 |
| 18 | 2.32 | 2.67 | 3.01 |
| 20 | 2.55 | 2.94 | 3.34 |
| 24 | 3.15 | 3.62 | 4.08 |

Hours listed above cover radiographic inspection of carbon steel butt welded joints by x-raying.

For radiographic inspection of mitre butt welds, add 50 percent to above hours.

For radiographic inspection of flange welds, add 100 percent to above hours.

For radiographic inspection of nozzle welds, add 200 percent to above hours.

# HYDROSTATIC TESTING PIPING SYSTEMS

HOURS REQUIRED PER LINEAR FOOT OF SIZES LISTED

| Pipe Size Inches | Standard Pipe & O.D. Sizes ⅜" Thick | Extra Heavy Pipe & O.D. Sizes ½" Thick | PIPE SCHEDULE NUMBERS | | | | | | | | |
|---|---|---|---|---|---|---|---|---|---|---|---|
| | | | 20 | 30 | 40 | 60 | 80 | 100 | 120 | 140 | 160 |
| 2" or Less | .014 | .016 | — | — | .014 | — | .016 | — | — | — | .025 |
| 2½ | .015 | .017 | — | — | .015 | — | .017 | — | — | — | .028 |
| 3 | .017 | .019 | — | — | .017 | — | .019 | — | — | — | .031 |
| 4 | .020 | .024 | — | — | .020 | — | .024 | — | .032 | — | .036 |
| 5 | .022 | .026 | — | — | .022 | — | .026 | — | .035 | — | .041 |
| 6 | .025 | .029 | — | — | .025 | — | .029 | — | .039 | — | .047 |
| 8 | .027 | .032 | .027 | .027 | .027 | .030 | .032 | .039 | .046 | .052 | .057 |
| 10 | .031 | .035 | .031 | .031 | .031 | .035 | .041 | .049 | .055 | .063 | .070 |
| 12 | .034 | .039 | .034 | .034 | .038 | .044 | .053 | .061 | .068 | .077 | .087 |
| 14 O.D. | .038 | .044 | .038 | .038 | .041 | .050 | .062 | .069 | .078 | .091 | .106 |
| 16 O.D. | .044 | .049 | .044 | .044 | .049 | .062 | .075 | .086 | .097 | .114 | .134 |
| 18 O.D. | .051 | .058 | .051 | .055 | .064 | .077 | .095 | .108 | .123 | .141 | .164 |
| 20 O.D. | .057 | .067 | .057 | .067 | .078 | .095 | .115 | .133 | .151 | .172 | .203 |
| 24 O.D. | .076 | .086 | .076 | .087 | .115 | .141 | .175 | .200 | .230 | .265 | .304 |

Above hours are average for testing completed systems for a maximum holding time of one hour and include time for the following operations, if required:

1) Place and remove blinds and blanks as required.
2) Opening and closing of valves.
3) Removal and replacement of valves, orifice plates, expansion joints, and short pieces of pipe as may be required.
4) Block-up and block removal of spring-supported or counterweight-supported lines.
5) Air purging of lines before hydro-test.
6) Soap testing joints where required.
7) Drain lines after testing.

If individual segments or spools are to be tested separately multiply above hours by a factor of 10.

# CONDUIT, BOXES AND FITTINGS

## Octagon, Square, Handy and Switch Boxes and Covers

HOURS REQUIRED PER UNITS LISTED

| | | PLATFORM WORK IN HEIGHT TO | | | |
|---|---|---|---|---|---|
| Item Description | Unit | 10′ | 15′ | 20′ | 25′ |
| 4-Inch Octagon Boxes and Covers: | | | | | |
| Octagon Box | Ea. | .32 | .33 | .34 | .35 |
| Octagon Box Extension | Ea. | .23 | .23 | .24 | .24 |
| Round Device Cover | Ea. | .22 | .23 | .24 | .25 |
| Round Blank Cover | Ea. | .15 | .15 | .16 | .17 |
| Round Cover with K.O. | Ea. | .15 | .15 | .16 | .17 |
| Round Switch or Receptical Cover | Ea. | .11 | .11 | .12 | .13 |
| Round Swivil Hanger Cover | Ea. | .11 | .11 | .12 | .13 |
| 4-Inch Square Boxes and Covers: | | | | | |
| Square Box | Ea. | .29 | .30 | .32 | .33 |
| Square Box Extension | Ea. | .22 | .23 | .24 | .25 |
| Square Cover—1 Device | Ea. | .22 | .23 | .24 | .25 |
| Square Cover—2 Device | Ea. | .22 | .23 | .24 | .25 |
| Square Offset Cover—1 Device | Ea. | .44 | .46 | .48 | .50 |
| Square Blank Cover | Ea. | .15 | .15 | .16 | .17 |
| Square Cover with K.O. | Ea. | .15 | .15 | .16 | .17 |
| Square Swivil Hanger Cover | Ea. | .16 | .16 | .17 | .18 |
| 4-11/16-Inch Square Boxes and Covers | | | | | |
| Square Box | Ea. | .29 | .30 | .32 | .33 |
| Square Box Extension | Ea. | .22 | .23 | .24 | .25 |
| Square Cover—1 Device | Ea. | .22 | .23 | .24 | .25 |
| Square Cover—2 Device | Ea. | .22 | .23 | .24 | .25 |
| Square Blank Cover | Ea. | .15 | .15 | .16 | .17 |
| Square Blank Cover with K.O. | Ea. | .15 | .15 | .16 | .17 |
| Handy Boxes: | | | | | |
| Handy Box | Ea. | .22 | .23 | .24 | .25 |
| Handy Box Extension | Ea. | .15 | .15 | .16 | .17 |
| Handy Box Cover | Ea. | .15 | .15 | .16 | .17 |
| Sectional Switch Box with K.O. | Ea. | .22 | .23 | .24 | .25 |
| Sectional Switch Box with Clamps | Ea. | .30 | .30 | .32 | .33 |

Above hours include checking out of barge storage, rigging, picking, setting, aligning and fastening to supports.

Time frames do not include installation or fastening of conduits or wire. See respective tables for these hours.

# CONDUIT, BOXES AND FITTINGS

## Gang Boxes and Covers

HOURS REQUIRED PER UNITS LISTED

| Item Description | Unit | PLATFORM WORK IN HEIGHT TO | | | |
|---|---|---|---|---|---|
| | | 10′ | 15′ | 20′ | 25′ |
| Gang Boxes: | | | | | |
| 2-Gang | Ea. | .41 | .42 | .45 | .46 |
| 3-Gang | Ea. | .51 | .53 | .56 | .58 |
| 4-Gang | Ea. | .65 | .68 | .70 | .72 |
| 5-Gang | Ea. | .80 | .83 | .85 | .87 |
| 6-Gang | Ea. | .96 | 1.00 | 1.02 | 1.04 |
| 7-Gang | Ea. | 1.15 | 1.20 | 1.22 | 1.25 |
| 8-Gang | Ea. | 1.38 | 1.44 | 1.46 | 1.50 |
| Gang Device Covers: | | | | | |
| 2-Gang Cover | Ea. | .15 | .15 | .16 | .17 |
| 3-Gang Cover | Ea. | .15 | .15 | .16 | .17 |
| 4-Gang Cover | Ea. | .15 | .15 | .16 | .17 |
| 5-Gang Cover | Ea. | .15 | .15 | .16 | .17 |
| 6-Gang Cover | Ea. | .23 | .23 | .24 | .25 |
| 7-Gang Cover | Ea. | .34 | .34 | .36 | .38 |
| 8-Gang Cover | Ea. | .51 | .51 | .54 | .57 |
| Gang Cover Surface Mounted Device: | | | | | |
| 2-Surface Mounted Device | Ea. | .18 | .18 | .19 | .20 |
| 3-Surface Mounted Device | Ea. | .18 | .18 | .19 | .20 |
| 4-Surface Mounted Device | Ea. | .18 | .18 | .19 | .20 |
| 5-Surface Mounted Device | Ea. | .18 | .18 | .19 | .20 |
| 6-Surface Mounted Device | Ea. | .28 | .28 | .29 | .30 |
| Gang Blank Covers: | | | | | |
| 2-Gang Blank Cover | Ea. | .15 | .15 | .16 | .17 |
| 3-Gang Blank Cover | Ea. | .15 | .15 | .16 | .17 |
| 4-Gang Blank Cover | Ea. | .15 | .15 | .16 | .17 |
| 5-Gang Blank Cover | Ea. | .15 | .15 | .16 | .17 |
| 6-Gang Blank Cover | Ea. | .23 | .23 | .24 | .25 |
| 7-Gang Blank Cover | Ea. | .34 | .34 | .36 | .38 |
| 8-Gang Blank Cover | Ea. | .51 | .51 | .54 | .57 |

Above hours include checking out of barge storage, rigging, picking, setting, aligning and fastening to supports.

Time frames do not include installation or fastening of conduits or wire. See respective tables for these hours.

# CONDUIT, BOXES AND FITTINGS

## Sheet Metal Boxes for Branch Rough-In

HOURS REQUIRED PER UNITS LISTED

| Item Description | Unit | PLATFORM WORK IN HEIGHT TO | | | |
|---|---|---|---|---|---|
| | | 10′ | 15′ | 20′ | 25′ |
| S.C. Pull Boxes: | | | | | |
| 4″ × 4″ × 4″ | Ea. | .68 | .71 | .73 | .75 |
| 4″ × 6″ × 4″ | Ea. | .68 | .71 | .73 | .75 |
| 6″ × 6″ × 4″ | Ea. | .68 | .71 | .73 | .75 |
| 6″ × 8″ × 4″ | Ea. | .72 | .76 | .78 | .80 |
| 8″ × 8″ × 4″ | Ea. | .72 | .76 | .78 | .80 |
| 8″ × 12″ × 4″ | Ea. | .90 | .95 | .98 | 1.01 |
| 12″ × 12″ × 4″ | Ea. | 1.08 | 1.13 | 1.16 | 1.19 |
| 12″ × 24″ × 4″ | Ea. | 1.71 | 1.80 | 1.85 | 1.91 |
| 12″ × 12″ × 6″ | Ea. | 1.35 | 1.42 | 1.46 | 1.50 |
| 12″ × 18″ × 6″ | Ea. | 1.53 | 1.61 | 1.66 | 1.71 |
| 12″ × 24″ × 6″ | Ea. | 1.71 | 1.80 | 1.85 | 1.91 |
| 18″ × 24″ × 6″ | Ea. | 2.25 | 2.36 | 2.43 | 2.50 |
| 18″ × 30″ × 6″ | Ea. | 2.48 | 2.60 | 2.68 | 2.76 |
| 24″ × 36″ × 6″ | Ea. | 3.38 | 3.55 | 3.66 | 3.77 |
| Hinge Cover Boxes: | | | | | |
| 6″ × 6″ × 4″ | Ea. | .68 | .71 | .73 | .75 |
| 6″ × 8″ × 4″ | Ea. | .72 | .76 | .78 | .80 |
| 8″ × 8″ × 4″ | Ea. | .72 | .76 | .78 | .80 |
| 8″ × 12″ × 4″ | Ea. | .90 | .95 | .98 | 1.01 |
| 12″ × 12″ × 4″ | Ea. | 1.08 | 1.13 | 1.16 | 1.19 |
| 12″ × 18″ × 4″ | Ea. | 1.53 | 1.61 | 1.66 | 1.71 |
| 12″ × 24″ × 4″ | Ea. | 1.71 | 1.80 | 1.85 | 1.91 |
| 12″ × 12″ × 6″ | Ea. | 1.35 | 1.42 | 1.46 | 1.50 |
| 12″ × 18″ × 6″ | Ea. | 1.53 | 1.61 | 1.66 | 1.71 |
| 12″ × 24″ × 6″ | Ea. | 1.71 | 1.80 | 1.85 | 1.91 |
| 18″ × 24″ × 6″ | Ea. | 2.25 | 2.36 | 2.43 | 2.50 |

Above hours include checking out of barge storage, rigging, picking, setting, aligning and fastening to supports.

Time frames do not inlcude installation or fastening of conduits or wire. See respective tables for these hours.

# CONDUIT, BOXES AND FITTINGS

## Installing Conduit

HOURS REQUIRED PER HUNDRED LINEAR FEET

| Item Description | Size In Inches | PLATFORM WORK IN HEIGHT TO | | | |
|---|---|---|---|---|---|
| | | 10′ | 15′ | 20′ | 25′ |
| Rigid Galvanized—Conduit ....... | ½ | 13.25 | 13.50 | 13.90 | 14.20 |
| Rigid Galvanized—Conduit ....... | ¾ | 16.30 | 16.65 | 17.15 | 17.50 |
| Rigid Galvanized—Conduit ....... | 1 | 23.50 | 24.00 | 24.70 | 25.20 |
| Rigid Galvanized—Conduit ....... | 1¼ | 24.25 | 24.75 | 25.50 | 26.00 |
| Rigid Galvanized—Conduit ....... | 1½ | 32.35 | 33.00 | 34.00 | 34.65 |
| Rigid Galvanized—Conduit ....... | 2 | 37.50 | 38.25 | 39.40 | 40.15 |
| Rigid Galvanized—Conduit ....... | 2½ | 44.10 | 45.00 | 46.35 | 47.15 |
| Rigid Galvanized—Conduit ....... | 3 | 58.80 | 60.00 | 61.80 | 63.00 |
| Rigid Galvanized—Conduit ....... | 3½ | 68.60 | 70.00 | 72.10 | 73.50 |
| Rigid Galvanized—Conduit ....... | 4 | 73.50 | 75.00 | 77.25 | 78.75 |
| Rigid Galvanized—Conduit ....... | 5 | 83.30 | 85.00 | 87.55 | 89.25 |
| Rigid Galvanized—Conduit ....... | 6 | 88.20 | 90.00 | 92.70 | 94.50 |

Above hours include checking out of barge storage, rigging, picking, setting, aligning and fastening conduit to supports.

For overhead work, add 20 percent to above hours.

In the case of parallel runs which are to be installed at the same time apply the following percentages of above hours for additional runs:

For Second Parallel Run—95%
For Third Parallel Run—91%
For Fourth Parallel Run—87%
For Fifth Parallel Run—84%
For Each Additional Run Above Five—80%

For installation of rigid aluminum conduit use 80 percent of above hours.

For installation of flexible steel conduit use 70 percent of above hours.

Above hours do not include cutting, reaming, threading and bending of conduit or handling of fittings and make-up of joints or connections. See respective tables for these hours.

# CONDUIT, BOXES AND FITTINGS

## Cutting, Reaming and Threading Conduit and Making-On of Joint

HOURS REQUIRED EACH

| Conduit Size Inches | CUT, REAM AND THREAD | | | MAKE-ON OR TREAD-ON PLATFORM WORK HEIGHT TO | | | |
|---|---|---|---|---|---|---|---|
| | Cutting Only | Ream & Thread | Combined Operation | 10′ | 15′ | 20′ | 25′ |
| ½ | .20 | .30 | .50 | .36 | .39 | .40 | .41 |
| ¾ | .20 | .30 | .50 | .36 | .40 | .41 | .42 |
| 1 | .20 | .30 | .50 | .60 | .65 | .68 | .69 |
| 1¼ | .23 | .35 | .58 | .60 | .65 | .68 | .71 |
| 1½ | .25 | .43 | .68 | .72 | .78 | .81 | .83 |
| 2 | .30 | .45 | .75 | .84 | .91 | .95 | .97 |
| 2½ | .30 | .45 | .75 | .90 | .97 | 1.00 | 1.04 |
| 3 | .40 | .50 | .90 | 1.08 | 1.17 | 1.22 | 1.24 |
| 3½ | .40 | .60 | 1.00 | 1.14 | 1.24 | 1.28 | 1.31 |
| 4 | .50 | .80 | 1.30 | 1.20 | 1.30 | 1.35 | 1.38 |
| 5 | .63 | 1.00 | 1.63 | 1.50 | 1.63 | 1.69 | 1.73 |
| 6 | .75 | 1.20 | 1.95 | 1.80 | 1.95 | 2.03 | 2.07 |

Above cutting, reaming and threading hours include measuring, checking, cutting with hand saw and threading and reaming by hand. If cutting is accomplished with power band saw and threading and reaming with power machine, use 75 percent of above hours.

Above make-on or tread-on hours include checking fittings out of barge storage, rigging, picking and setting on deck, applying joint sealer to threads and single-hub make-on. Hours are average for one tread-on only for any type fitting. It must be remembered that a coupling, a 90° ell, or a 45° ell has two tread-ons; a cap, one tread-on, etc.

# CONDUIT BENDING

HOURS REQUIRED PER BEND

| Conduit Size Inches | NUMBER AND TYPE OF BENDS | | | | |
|---|---|---|---|---|---|
| | 1, 2, 3 & 4 | 5 | 6 | 7 | 8 |
| ½ | .30 | .68 | .41 | .40 | .68 |
| ¾ | .35 | .77 | .54 | .45 | .77 |
| 1 | .55 | 1.11 | .78 | .60 | 1.11 |
| 1¼ | .62 | 2.21 | 1.55 | .70 | 2.21 |
| 1½ | .80 | 2.47 | 1.73 | .90 | 2.47 |
| 2 | 1.00 | 2.89 | 2.00 | 1.20 | 2.89 |
| 2½ | 1.25 | 3.32 | 2.32 | 1.60 | 3.32 |
| 3 | 1.50 | 3.74 | 2.62 | 1.85 | 3.74 |
| 3½ | 1.80 | 4.42 | 3.10 | 2.00 | 4.42 |
| 4 | 2.50 | 4.85 | 3.40 | 2.20 | 4.85 |
| 5 | 3.13 | — | — | — | — |
| 6 | 3.75 | — | — | — | — |

Hours include use of standard lengths or pieces of conduit in proximity of final installation. Bends are made by hand as single or multiple operations on standard portable equipment. Hours for ½-inch through 1-inch bends are for general use hickie.

Hours do not include cutting, reaming, threading, make-ons or installation. See respective tables for these time frames.

# STANDARD TYPES OF BENDS

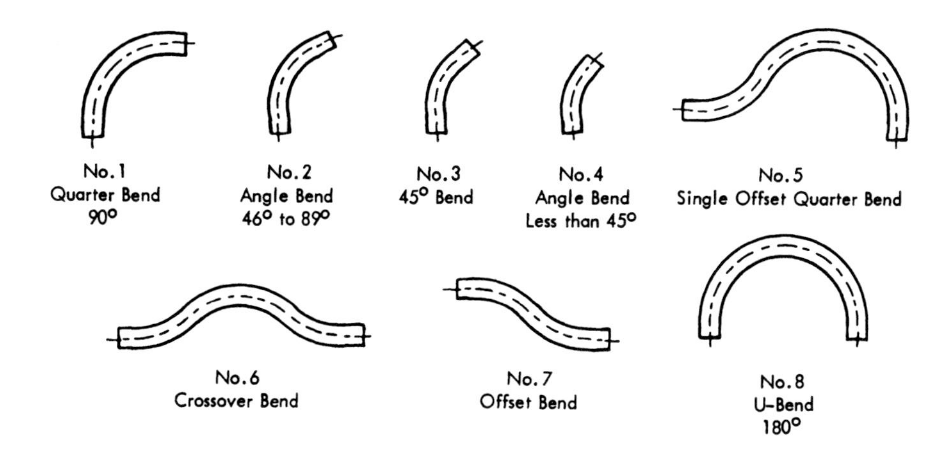

# SERVICE AND FEEDER WIRING
# WIRE PULLING—SIMPLE LAY-OUT

**Stranded or Solid Wire**
**Rubber or Thermoplastic Insulated Covered**

HOURS FOR NUMBER OF WIRES AND RUNS AS LISTED

| Wire Size | 50′ CONDUIT RUNS | | | 100′ CONDUIT RUNS | | | 200′ CONDUIT RUNS | | |
|---|---|---|---|---|---|---|---|---|---|
| | 2-wire | 3-wire | 4-wire | 2-wire | 3-wire | 4-wire | 2-wire | 3-wire | 4-wire |
| #18 | .54 | .72 | .99 | 1.04 | 1.50 | 1.89 | 2.02 | 2.87 | 3.67 |
| #16 | .66 | .88 | 1.21 | 1.27 | 1.83 | 2.31 | 2.47 | 3.51 | 4.49 |
| #14 | .81 | 1.08 | 1.48 | 1.54 | 2.22 | 2.80 | 3.00 | 4.26 | 5.44 |
| #12 | .99 | 1.57 | 1.80 | 1.98 | 2.70 | 3.40 | 3.64 | 5.22 | 6.64 |
| #10 | 1.39 | 1.98 | 2.52 | 2.64 | 3.78 | 4.80 | 5.12 | 7.38 | 9.28 |
| #8 | 1.68 | 2.40 | 3.04 | 3.20 | 4.56 | 5.80 | 6.20 | 8.88 | 11.28 |
| #6 | 1.74 | 2.49 | 3.16 | 3.30 | 4.74 | 6.00 | 6.40 | 9.18 | 11.68 |
| #4 | 2.09 | 2.97 | 3.76 | 3.98 | 5.67 | 7.20 | 7.72 | 10.98 | 13.92 |
| #2 | 2.32 | 3.30 | 4.16 | 4.42 | 6.27 | 8.00 | 8.56 | 12.24 | 15.44 |
| #1 | 2.66 | 3.78 | 4.80 | 5.06 | 7.23 | 9.20 | 9.84 | 14.10 | 17.84 |
| #1/0 | 3.37 | 4.80 | 6.08 | 6.42 | 9.15 | 11.60 | 12.44 | 17.76 | 22.48 |
| #2/0 | 3.70 | 5.28 | 6.68 | 7.06 | 10.05 | 12.76 | 13.68 | 19.56 | 24.72 |
| #3/0 | 4.40 | 6.27 | 7.96 | 8.38 | 11.94 | 15.16 | 16.28 | 23.22 | 29.44 |
| #4/0 | 4.98 | 7.11 | 9.00 | 9.48 | 13.53 | 17.16 | 18.40 | 26.28 | 33.36 |
| 250 MCM | 5.21 | 7.44 | 9.40 | 9.92 | 14.13 | 17.96 | 19.68 | 28.04 | 34.80 |
| 300 MCM | 5.57 | 7.96 | 10.06 | 10.61 | 15.12 | 19.22 | 21.06 | 30.00 | 37.24 |
| 350 MCM | 5.79 | 8.28 | 10.48 | 11.04 | 15.72 | 19.96 | 21.40 | 30.60 | 38.72 |
| 400 MCM | 6.20 | 8.86 | 11.21 | 11.81 | 16.82 | 21.36 | 22.90 | 32.74 | 41.43 |
| 500 MCM | 6.72 | 9.60 | 12.16 | 12.80 | 18.24 | 23.16 | 24.84 | 35.46 | 44.88 |
| 600 MCM | 7.19 | 10.26 | 13.00 | 13.68 | 19.50 | 24.76 | 26.56 | 37.86 | 48.00 |
| 750 MCM | 9.26 | 13.23 | 16.76 | 17.64 | 25.14 | 31.92 | 34.20 | 48.90 | 61.92 |
| 1,000 MCM | 11.12 | 15.87 | 20.08 | 21.16 | 30.15 | 38.32 | 41.08 | 58.68 | 74.32 |

A simple lay-out is one that presents normal working conditions and conduit runs free from obstacles such as fittings or pull boxes at right angle turns.

Above hours include checking out of barge storage, measuring, cutting, group and prepare ends, fish and pull lines, connect and pull cable and "ringing out" or identifying the circuit conductors.

If duplicate pulls are to be made, deduct the following percentages from the above hours to compensate for equipment set-up time:

#18—#8—4%
#6—#4—5%
#2—#1—7%
#1/0—1,000 MCM—9%

# SERVICE AND FEEDER WIRING
# WIRE PULLING—COMPLEX LAY-OUT

**Stranded or Solid Wire**
**Rubber or Thermoplastic Insulated Covered**

HOURS FOR NUMBER OF WIRES AND RUNS AS LISTED

| Wire Size | 50′ CONDUIT RUNS | | | 100′ CONDUIT RUNS | | | 200′ CONDUIT RUNS | | |
|---|---|---|---|---|---|---|---|---|---|
| | 2-wire | 3-wire | 4-wire | 2-wire | 3-wire | 4-wire | 2-wire | 3-wire | 4-wire |
| # 18 | .78 | 1.11 | 1.41 | 1.48 | 2.12 | 2.69 | 2.88 | 4.12 | 5.23 |
| # 16 | .95 | 1.35 | 1.71 | 1.80 | 2.57 | 3.26 | 3.49 | 4.99 | 6.34 |
| # 14 | 1.16 | 1.65 | 2.10 | 2.20 | 3.15 | 4.00 | 4.28 | 6.12 | 7.76 |
| # 12 | 1.41 | 2.01 | 2.56 | 2.68 | 3.84 | 4.80 | 5.20 | 7.44 | 9.44 |
| # 10 | 1.98 | 2.82 | 3.60 | 3.78 | 5.40 | 6.84 | 6.52 | 10.50 | 13.28 |
| # 8 | 2.40 | 3.42 | 4.34 | 4.58 | 6.54 | 8.28 | 8.88 | 12.61 | 16.08 |
| # 6 | 2.48 | 3.54 | 4.50 | 4.72 | 6.75 | 9.56 | 9.66 | 13.08 | 16.64 |
| # 4 | 2.98 | 4.22 | 5.40 | 5.68 | 8.10 | 10.28 | 11.00 | 15.72 | 19.92 |
| # 2 | 3.31 | 4.71 | 5.98 | 6.30 | 9.00 | 11.40 | 12.24 | 17.46 | 22.08 |
| # 1 | 3.80 | 5.43 | 6.88 | 7.24 | 10.35 | 13.12 | 14.04 | 20.10 | 25.44 |
| # 1/0 | 4.81 | 6.87 | 9.70 | 9.16 | 13.08 | 16.56 | 17.76 | 25.38 | 32.16 |
| # 2/0 | 5.29 | 7.56 | 9.58 | 10.08 | 14.40 | 18.24 | 19.56 | 27.96 | 35.36 |
| # 3/0 | 6.29 | 8.97 | 11.38 | 11.98 | 17.10 | 21.68 | 23.24 | 33.18 | 42.08 |
| # 4/0 | 7.11 | 10.14 | 12.88 | 13.54 | 19.35 | 24.52 | 26.28 | 37.56 | 47.60 |
| 250 MCM | 7.44 | 10.62 | 13.46 | 14.18 | 20.25 | 25.64 | 27.52 | 39.30 | 49.76 |
| 300 MCM | 7.96 | 11.36 | 14.40 | 15.17 | 21.67 | 27.43 | 29.45 | 42.05 | 53.24 |
| 350 MCM | 8.27 | 11.82 | 14.98 | 15.76 | 22.50 | 28.52 | 30.56 | 43.68 | 55.36 |
| 400 MCM | 8.85 | 12.65 | 16.03 | 16.86 | 24.08 | 30.52 | 32.70 | 46.74 | 59.24 |
| 500 MCM | 9.60 | 12.71 | 17.36 | 18.28 | 26.10 | 33.08 | 35.48 | 50.64 | 64.16 |
| 600 MCM | 10.26 | 14.64 | 18.56 | 19.54 | 27.90 | 35.36 | 37.92 | 54.12 | 68.56 |
| 750 MCM | 13.23 | 18.90 | 23.94 | 25.20 | 36.00 | 45.60 | 48.88 | 69.84 | 88.48 |
| 1,000 MCM | 15.88 | 22.68 | 28.72 | 30.24 | 43.20 | 54.72 | 58.68 | 83.82 | 106.16 |

A complex lay-out is one that requires pulling cable for various locations under adverse working conditions or through pull boxes at right angles. There are many and varying degrees of complexity to be considered.

Above hours include checking out of barge storage, measuring, cutting, group and prepare ends, fish and pull lines, connect and pull cable, and "ringing out" or identifying the circuit conductors.

If duplicate pulls are to be made, deduct the following percentages from the above hours to compensate for equipment set-up time:

# 18—# 8—4%
# 6—# 4—5%
# 2—# 1—7%
# 1/0—1,000 MCM—9%

# CABLE INSTALLATION

## Flexible Metallic Armored Cable (BX and BXL)

HOURS REQUIRED FOR NUMBER OF CONDUCTORS AND RUNS AS LISTED

| Number & Size Conductors | 50′ RUNS | | 100′ RUNS | | 200′ RUNS | |
|---|---|---|---|---|---|---|
| | Concealed | Exposed | Concealed | Exposed | Concealed | Exposed |
| BX Cable—Copper Conductor: | | | | | | |
| 2/ # 14 | 2.75 | 2.36 | 5.25 | 4.50 | 9.98 | 8.56 |
| 3/ # 14 | 2.99 | 2.60 | 5.70 | 4.95 | 10.84 | 9.40 |
| 4/ # 14 | 3.11 | 2.72 | 5.95 | 5.20 | 11.25 | 9.80 |
| 2/ # 12 | 2.99 | 2.60 | 5.70 | 4.95 | 10.84 | 9.40 |
| 3/ # 12 | 3.54 | 2.75 | 6.75 | 5.25 | 12.82 | 9.98 |
| 4/ # 12 | | | | | | |
| 2/ # 10 | 3.94 | 3.94 | 7.50 | 7.50 | 14.26 | 14.26 |
| 3/ # 10 | 4.72 | 4.41 | 9.00 | 8.40 | 17.10 | 15.96 |
| 4/ # 10 | 5.10 | 4.65 | 9.75 | 8.95 | 18.52 | 16.81 |
| 2/ # 8 | 5.12 | 4.92 | 9.75 | 9.38 | 18.52 | 17.82 |
| 3/ # 8 | 5.90 | 5.51 | 11.25 | 10.50 | 21.38 | 19.96 |
| 4/ # 8 | 6.30 | 5.80 | 12.00 | 11.06 | 22.80 | 21.00 |
| BXL Cable—Copper Conductor: | | | | | | |
| 2/ # 14 | 2.99 | 2.60 | 5.70 | 4.95 | 10.84 | 9.40 |
| 3/ # 14 | 3.15 | 2.75 | 6.00 | 5.25 | 11.40 | 9.98 |
| 2/ # 12 | 3.15 | 2.75 | 6.00 | 5.25 | 11.40 | 9.98 |
| 3/ # 12 | 3.74 | 3.35 | 7.13 | 6.38 | 13.54 | 12.12 |
| 2/ # 10 | 4.33 | 3.94 | 8.25 | 7.50 | 15.68 | 14.26 |
| 3/ # 10 | 5.12 | 4.72 | 9.75 | 9.00 | 18.52 | 17.10 |
| 2/ # 8 | 5.32 | 5.12 | 10.13 | 9.75 | 19.24 | 18.52 |
| 3/ # 8 | 6.30 | 5.90 | 12.00 | 11.25 | 22.80 | 21.38 |
| Armored Single Wire—Copper Conductor: | | | | | | |
| 1/ # 8 | 2.00 | 1.70 | 3.82 | 3.44 | 8.48 | 7.63 |
| 1/ # 6 | 2.50 | 2.13 | 4.78 | 4.30 | 10.61 | 9.55 |
| 1/ # 4 | 3.00 | 3.53 | 5.73 | 5.16 | 12.72 | 11.45 |

Above hours include checking out of barge storage, rigging, picking, setting on platform deck, "boring out," pulling in cable, "ringing out" or identifying and fastening and securing to structure.

Hours are average for length of runs as shown in above table regardless of height.

Hours do not include splicing or connecting to switches or receptacles. See respective tables for these time requirements.

# CABLE CONNECTORS AND LUGS

HOURS REQUIRED EACH

| Wire Size | TYPE OF CONNECTOR OR LUG | | | |
|---|---|---|---|---|
| | A | B | C | D |
| 6 | .54 | .45 | .27 | .23 |
| 4 | .63 | .54 | .27 | .23 |
| 2 | .72 | .63 | .27 | .23 |
| 1 | 1.04 | .90 | .54 | .45 |
| 1/0 | 1.13 | 1.04 | .68 | .59 |
| 2/0 | 1.35 | 1.13 | .81 | .68 |
| 3/0 | 1.58 | 1.35 | .99 | .81 |
| 4/0 | 1.80 | 1.58 | 1.13 | .90 |
| 250 MCM | 2.25 | 1.80 | 1.40 | 1.13 |
| 300 MCM | 2.48 | 2.03 | 1.58 | 1.35 |
| 350 MCM | 2.70 | 2.48 | 1.71 | 1.44 |
| 400 MCM | 3.15 | 2.70 | 1.80 | 1.58 |
| 500 MCM | 3.60 | 3.15 | 2.07 | 1.80 |
| 750 MCM | 4.50 | 3.60 | 3.15 | 2.48 |
| 1,000 MCM | 5.40 | 4.50 | 3.60 | 2.93 |

Code:

A = U-Bolt Connector—Tape Wrapped
B = Two-Way Compression Connector—Tape Wrapped
C = Bolt Lugs
D = Compression Lugs

# CABLE VERTICAL RISER SUPPORTS

HOURS REQUIRED EACH

| Size Inches | HEIGHT TO | | | |
|---|---|---|---|---|
| | 10′ | 15′ | 20′ | 25′ |
| 1¼ | .77 | .79 | .81 | .83 |
| 1½ | .99 | 1.02 | 1.05 | 1.08 |
| 2 | 1.49 | 1.53 | 1.58 | 1.63 |
| 2½ | 1.98 | 2.04 | 2.10 | 2.16 |
| 3 | 2.48 | 2.55 | 2.63 | 2.71 |
| 3½ | 2.97 | 3.06 | 3.15 | 3.24 |
| 4 | 3.83 | 3.95 | 4.07 | 4.19 |
| 5 | 4.73 | 4.87 | 5.02 | 5.18 |
| 6 | 5.85 | 6.03 | 6.21 | 6.40 |

Hours include checking out of barge storage, rigging, picking setting on platform and installing.

Hours do not include cable installation. See page 262 for this time requirement.

# SWITCHES AND PLATES

HOURS REQUIRED EACH

| Type | Item Description | Device Only | Standard Plates Only | Complete With Plates |
|---|---|---|---|---|
| 10A | Single-Pole Switch | .27 | .15 | .42 |
| 10A | Double-Pole Switch | .75 | .15 | .90 |
| 10A | Three-Way Switch | .53 | .15 | .68 |
| 10A | Four-Way Switch | .75 | .15 | .90 |
| 10A | Double-Pole Three-Point Switch | .60 | .15 | .75 |
| 10A | Double-Pole Four-Point Switch | .60 | .15 | .75 |
| 20A | Single-Pole Switch | .45 | .15 | .60 |
| 20A | Double-Pole Switch | 1.05 | .15 | 1.20 |
| 20A | Three-Way Switch | .68 | .15 | .83 |
| 20A | Four-Way Switch | 1.05 | .15 | 1.20 |
| 30A | Single-Pole Switch | .60 | .15 | .75 |
| 30A | Double-Pole Switch | 1.13 | .15 | 1.28 |
| 30A | Three-Way Switch | .90 | .15 | 1.05 |
| 30A | Four-Way Switch | 1.13 | .15 | 1.28 |
| 10A | Single-Pole Ceiling Pull Switch | — | — | .75 |
| 10A | Three-Way Ceiling Pull Switch | — | — | 1.13 |
| — | Pendant Switch and Cord | — | — | .75 |
| — | Pendant Switch and Heavy Duty Cord | — | — | 1.05 |
| — | Canopy Switch (Assembled with Fixture) | — | — | .23 |
| — | Door Switch and Box | — | — | 2.25 |
| 10A | Mark Time Switch | — | — | .68 |
| 20A | Mark Time Switch | — | — | .90 |

Above hours include checking out of barge storage, rigging, picking, setting on platforms and complete installation including make-up of connection.

"Device Only" column includes installation of switch and making up.

"Standard Plates Only" includes a separate operation for installing plate.

"Complete With Plates" includes hours for the installation of device and plate as one operation.

# RECEPTACLES, PLATES, AND MISCELLANEOUS OUTLETS AND COVERS

HOURS REQUIRED EACH

| Type | Item Description | Device Only | Standard Plates Only | Complete With Plates |
|---|---|---|---|---|
| 10A | Duplex Receptacle | .27 | .15 | .42 |
| 20A | Two-Wire Receptacle | .42 | .15 | .57 |
| 30A | Two-Wire Receptacle | .68 | .15 | .83 |
| 10A | Three-Wire Receptacle | .45 | .15 | .60 |
| 20A | Three-Wire Receptacle | .57 | .15 | .72 |
| 30A | Three-Wire Receptacle | .75 | .15 | .90 |
| 20A | Three-Wire Ground-Type Receptacle | .53 | .15 | .68 |
| 30A | Three-Wire Ground-Type Receptacle | .68 | .15 | .83 |
| 50A | Three-Wire Ground-Type Receptacle | 1.05 | .15 | 1.20 |
| 20A | Four-Wire Ground-Type Receptacle | .68 | .15 | .83 |
| 30A | Four-Wire Ground-Type Receptacle | .83 | .15 | .98 |
| 50A | Four-Wire Ground-Type Receptacle | 1.20 | .15 | 1.35 |
| 10A | Weatherproof Receptacle | — | — | .38 |
| 10A | Clock Hanger Outlet | .53 | .23 | .76 |
| 10A | Fan Hanger Outlet | .83 | .23 | 1.06 |
| — | Pilot Light Receptacle | .60 | .23 | .83 |
| — | Cover Socket—Ceiling | — | — | .60 |
| — | Floor Outlet with Plate—One Gang | — | — | 2.25 |
| — | Floor Outlet W.P.—Gang Type (Per Gang) | — | — | 1.88 |
| 15A | Receptacle in Floor Box | .68 | — | — |
| — | Receptacle Stand—Floor Box | .60 | — | — |
| — | Bell Nozzle—Floor Box | .53 | — | — |
| — | Switch and Receptacle Plate (Special) | — | .30 | — |
| — | Radio Outlet | — | — | .83 |

Above hours include checking out of barge storage, rigging, picking, setting on platform and complete installation, including make-up of connection.

"Device Only" column includes installation of receptacle or outlet and make-up operation.

"Standard Plates Only" includes a separate operation for installing plate only.

"Complete With Plates" includes hours for the installation of receptacles and plates as one operation.

# STANDARD PANELS AND CABINETS

HOURS REQUIRED FOR PANELBOARD AND CABINET

| Number of 30-Amp Two-Wire Branch Circuits | HOURS REQUIRED | |
|---|---|---|
| | Flush-Mounted | Surface-Mounted |
| 4 | 4.95 | 4.50 |
| 6 | 7.50 | 6.75 |
| 8 | 9.00 | 8.10 |
| 10 | 10.50 | 9.45 |
| 12 | 13.50 | 12.15 |
| 14 | 15.00 | 14.10 |
| 16 | 16.50 | 14.86 |
| 18 | 18.00 | 16.20 |
| 20 | 19.50 | 17.55 |
| 22 | 21.00 | 18.90 |
| 24 | 22.50 | 20.25 |
| 26 | 24.00 | 21.60 |
| 28 | 25.50 | 22.95 |
| 30 | 27.00 | 24.30 |
| 32 | 28.50 | 25.65 |
| 34 | 33.00 | 29.70 |
| 36 | 34.50 | 31.05 |
| 38 | 36.00 | 32.40 |
| 40 | 37.90 | 34.10 |
| 42 | 39.79 | 35.81 |

Hours are for the installation of standard panelboards and cabinets having fuses only or fuses and switches in the branches and having mains with lugs only, main fuses, main switches or main switches and fuses or circuit breaker.

Time has been allowed for the removal of insides and placing in temporary barge storage, rigging picking and setting box on platform, punching necessary knock-outs, situating and installing box, returning to barge storage and pick-up insides and installing in cabinet, installing main pipe terminals, and installing cover plate.

Hours do not include installation of fasteners, supports, sub-feeder terminals or sub-feeder pipe entrances. See respective tables for these time frames.

# 250- AND 600-VOLT PANELS AND CABINETS

**For Power and Lighting Distribution**

HOURS REQUIRED PER TERMINAL

| Branch Circuit Terminal Size Amperes | 250 VOLTS | | | | 600 VOLTS | | | |
|---|---|---|---|---|---|---|---|---|
| | Under 5 Circuits | | Over 5 Circuits | | Under 5 Circuits | | Over 5 Circuits | |
| | Soldered | Solderless | Soldered | Solderless | Soldered | Solderless | Soldered | Solderless |
| Standard Panels | | | | | | | | |
| 30 | .66 | .60 | .53 | .48 | .72 | .66 | .57 | .53 |
| 60 | .98 | .87 | .78 | .71 | 1.05 | .95 | .86 | .78 |
| 100 | 1.61 | 1.46 | 1.29 | 1.16 | 1.80 | 1.62 | 1.43 | 1.29 |
| 200 | 2.25 | 2.03 | 1.83 | 1.65 | 2.48 | 2.25 | 2.01 | 1.80 |
| 400 | 3.53 | 3.15 | 2.81 | 2.55 | 3.90 | 3.53 | 3.08 | 2.78 |
| 600 | 4.36 | 3.90 | 3.45 | 3.15 | 4.73 | 4.28 | 3.75 | 3.38 |
| 800 | 6.38 | 5.70 | 5.10 | 4.50 | 7.05 | 6.38 | 5.63 | 5.10 |
| Pull-Out Panels | | | | | | | | |
| 30 | .72 | .66 | .57 | .53 | .80 | .72 | .63 | .57 |
| 60 | 1.08 | .96 | .86 | .78 | 1.20 | 1.08 | .95 | .86 |
| 100 | 1.79 | 1.61 | 1.43 | 1.26 | 1.98 | 1.77 | 1.58 | 1.43 |
| 200 | 2.48 | 2.25 | 2.01 | 1.80 | 2.78 | 2.48 | 2.21 | 2.03 |
| 400 | 3.75 | 3.38 | 3.08 | 2.78 | 4.80 | 4.36 | 3.83 | 3.45 |
| 600 | 4.73 | 4.28 | 3.75 | 3.38 | 5.25 | 4.65 | 4.13 | 3.75 |
| 800 | 7.05 | 6.38 | 5.63 | 5.00 | 7.80 | 7.05 | 6.23 | 5.63 |
| Circuit-Breaker Panels | | | | | | | | |
| 30 | .83 | .75 | .66 | .60 | .90 | .83 | .72 | .65 |
| 60 | 1.20 | 1.08 | .98 | .87 | 1.36 | 1.20 | 1.08 | .98 |
| 100 | 2.03 | 1.83 | 1.62 | 1.46 | 2.25 | 2.03 | 1.80 | 1.62 |
| 200 | 2.86 | 2.55 | 2.30 | 2.06 | 3.18 | 2.86 | 2.55 | 2.30 |
| 400 | 4.50 | 4.05 | 3.53 | 3.15 | 4.88 | 4.36 | 3.87 | 3.48 |
| 600 | 5.40 | 4.88 | 4.36 | 3.90 | 6.00 | 5.40 | 4.80 | 4.33 |
| 800 | 7.95 | 7.13 | 6.38 | 5.70 | 8.63 | 7.80 | 6.90 | 6.23 |

Above hours include the removal of insides and placing in temporary barge storage, rigging, picking and setting cabinet on platform, punching necessary knock-outs, situating and installing cabinet, returning to barge storage and pick-up insides, and installing in cabinet and installing cover plates.

Hours do not include installation of fasteners, supports, sub-feeder terminals or sub-feeder pipe entrances. See respective tables for these time frames.

# 250- AND 600-VOLT SAFETY SWITCHES

HOURS REQUIRED EACH

| Amperes | Poles | 250 VOLTS | | 600 VOLTS | |
|---|---|---|---|---|---|
| | | Soldered | Solderless | Soldered | Solderless |
| 30 | 2 | 4.28 | 3.86 | 4.73 | 4.28 |
| 30 | 3 | 4.43 | 3.98 | 4.88 | 4.43 |
| 30 | 4 | 4.73 | 4.28 | 5.25 | 4.73 |
| 60 | 2 | 5.10 | 4.59 | 5.63 | 5.10 |
| 60 | 3 | 6.08 | 5.46 | 6.75 | 6.08 |
| 60 | 4 | 7.13 | 6.41 | 7.88 | 7.13 |
| 100 | 2 | 6.75 | 6.00 | 7.50 | 6.75 |
| 100 | 3 | 8.10 | 7.36 | 9.00 | 8.10 |
| 100 | 4 | 9.45 | 8.55 | 10.50 | 9.45 |
| 200 | 2 | 12.00 | 11.40 | 13.95 | 12.60 |
| 200 | 3 | 15.00 | 13.05 | 16.20 | 14.55 |
| 200 | 4 | 16.50 | 15.00 | 18.45 | 16.65 |
| 400 | 2 | 18.00 | 16.50 | 21.00 | 18.00 |
| 400 | 3 | 22.50 | 21.00 | 25.50 | 22.50 |
| 400 | 4 | 27.00 | 24.00 | 30.00 | 27.00 |
| 600 | 2 | 20.80 | 18.20 | 23.40 | 20.80 |
| 600 | 3 | 28.60 | 26.00 | 31.20 | 28.60 |
| 600 | 4 | 32.50 | 28.60 | 36.40 | 32.50 |
| 800 | 2 | 33.80 | 29.90 | 36.40 | 33.80 |
| 800 | 3 | 42.90 | 39.00 | 46.80 | 42.90 |
| 800 | 4 | 49.40 | 44.20 | 54.60 | 49.40 |

Hours include checking out of barge storage, rigging and hoisting to platform, mounting and connecting, including two-pipe entrances for fusible or non-fusible externally operated switches.

Hours do not include conduit runs, wire pulling or box installation. See respective tables for these time frames.

# HANGERS AND FASTENERS

HOURS REQUIRED PER HUNDRED

| Size Inches | ONE HOLE STRAP TYPE | | | | SPLIT PIPE RINGS & SOCKETS | | | | PIPE RISER CLAMPS | | | |
|---|---|---|---|---|---|---|---|---|---|---|---|---|
| | HEIGHT TO: | | | | HEIGHT TO: | | | | HEIGHT TO: | | | |
| | 10′ | 15′ | 20′ | 25′ | 10′ | 15′ | 20′ | 25′ | 10′ | 15′ | 20′ | 25′ |
| ⅜ | 1.37 | 1.40 | 1.44 | 1.47 | — | — | — | — | — | — | — | — |
| ½ | 1.37 | 1.40 | 1.44 | 1.47 | — | — | — | — | — | — | — | — |
| ¾ | 1.37 | 1.40 | 1.44 | 1.47 | — | — | — | — | — | — | — | — |
| 1 | 1.86 | 1.90 | 1.95 | 2.00 | — | — | — | — | — | — | — | — |
| 1¼ | 2.74 | 2.80 | 2.88 | 2.94 | 17.46 | 18.00 | 18.55 | 18.90 | 48.01 | 49.50 | 51.00 | 51.98 |
| 1½ | 2.74 | 2.80 | 2.88 | 2.94 | 26.19 | 27.00 | 27.80 | 28.35 | 53.84 | 55.50 | 57.15 | 58.28 |
| 2 | 4.12 | 4.20 | 4.33 | 4.41 | 33.46 | 34.50 | 35.55 | 36.20 | 58.20 | 60.00 | 61.80 | 63.00 |
| 2½ | 4.12 | 4.20 | 4.33 | 4.41 | 40.74 | 42.00 | 43.25 | 44.10 | 62.57 | 64.50 | 66.45 | 67.72 |
| 3 | 6.86 | 7.00 | 7.21 | 7.35 | 48.00 | 49.50 | 51.00 | 51.98 | 68.39 | 70.50 | 72.60 | 74.00 |
| 3½ | 8.43 | 8.60 | 8.86 | 9.03 | 62.56 | 64.50 | 68.45 | 67.70 | 77.12 | 79.50 | 81.90 | 83.47 |
| 4 | 9.80 | 10.00 | 10.30 | 10.50 | 69.85 | 72.00 | 74.15 | 75.60 | 87.30 | 90.00 | 92.70 | 94.50 |

| Size Inches | Item Description | HEIGHT TO: | | | |
|---|---|---|---|---|---|
| | | 10′ | 15′ | 20′ | 25′ |
| — | Beam Clamps | 26.30 | 27.00 | 27.80 | 28.35 |
| ¼ | Rod-Size Expansion Anchors | 29.30 | 30.00 | 30.90 | 31.50 |
| ⅜ | Rod-Size Expansion Anchors | 47.50 | 48.00 | 49.45 | 50.40 |
| ½ | Rod-Size Expansion Anchors | 56.50 | 57.00 | 58.70 | 59.85 |
| — | Concrete Inserts ⅜″ or ½″ Nuts | 36.75 | 37.50 | 38.65 | 39.40 |
| — | Ceiling Flanges and Sockets | 29.30 | 30.00 | 30.90 | 31.50 |

Hours include checking out of barge storage, rigging and hoisting to site of erection and installing.

Hours do not include installation of electrical devices. See respective tables for these time frames.

# MISCELLANEOUS FASTENERS

HOURS REQUIRED PER EACH OR OPERATION

| Bolt Size Inches | Lead Expansion Anchors | Toggle Bolts | Wooden and Lag Screws | Mach. Screws In Steel Drill & Tap | Mach. Bolt In Steel Av. ⅜" Thick |
|---|---|---|---|---|---|
| ⅛" | — | .11 | — | .32 | .24 |
| 3/16" | .14 | .12 | — | .39 | .27 |
| ¼" | .15 | .14 | — | .42 | .30 |
| 5/16" | .18 | — | — | .48 | .34 |
| ⅜" | .25 | .18 | — | .58 | .38 |
| 7/16" | .28 | — | — | .65 | .44 |
| ½" | .28 | — | — | — | — |
| ⅝" | .38 | — | — | — | — |
| #10 × 1" | — | — | .03 | — | — |
| #12 × 1¼" | — | — | .03 | — | — |
| ¼" × 1½" | — | — | .05 | — | — |
| ⅜" × 2" | — | — | .08 | — | — |
| ½" × 2½" | — | — | .11 | — | — |

Above hours include checking out of barge storage, rigging and hoisting to erection site, fabricating hole with power tool when required, and erection of anchor bolt or screw.

Hours are average for heights to 25 feet.

Hours do not include straps or installation of other supports. See respective tables for these time frames.

# MOTOR STARTING SWITCHES AND DIAL-TYPE SPEED-REGULATING RHEOSTATS

HOURS REQUIRED EACH

| MOTOR STARTING SWITCHES FOR 30-AMP AC MOTORS | | | DIAL-TYPE SPEED-REGULATING RHEOSTATS FOR 220-VOLT 3-PHASE SLIP-RING AC INDUCTION MOTORS | |
|---|---|---|---|---|
| Type Switch & Motor | Soldered | Solderless | Motor HP | Mounting |
| 2-Pole, Single-Phase | 5.63 | 5.10 | 1 | 6.60 |
| 3-Pole, 3-Phase or | | | 2-3 | 7.50 |
| 3-Wire, 2-Phase | 6.38 | 5.78 | 5 | 8.40 |
| 4-Pole, 4-Wire, | | | 7½-10 | 10.80 |
| 2-Phase | 6.75 | 6.08 | 15 | 12.60 |

HOURS REQUIRED EACH

| 30-Amp AC Magnetic Switches | | 3-Pole 220-Volt AC Magnetic Switches | |
|---|---|---|---|
| Number of Poles | Mounting | Capacity of Switch Amperes | Mounting |
| 3 | 7.50 | 15 | 6.00 |
| 4 | 8.70 | 75 | 9.45 |
| — | — | 150 | 12.00 |
| — | — | 300 | 15.00 |

Above hours include:

*Motor Starting Switches for 30-Amp AC Motors*—Making connection at motor and testing for direction of rotation for 30-Amp., motor-starting switches.

*Dial-Type Regulating Rheostats*—Mounting switch and rheostat and making connection at motor.

*30-Amp AC Magnetic Switches*—Mounting and connecting with thermal cutouts or relays used as starters for small squirrel-cage motors, for mounting and connecting push-button control station and making connection at the motor and testing.

*3-Pole 220-Volt AC Magnetic Switches*—Mounting and connecting push-button control station.

Hours do not include installation of conduit, pulling of wire or mounting of motors. See other tables for these time frames.

# STARTING COMPENSATORS

**For 3-Phase Squirrel Cage Induction AC Motors**

| Voltage | Motor Horsepower | Mounting |
|---|---|---|
| 220 | 7½-10 | 8.25 |
| 220 | 15 | 10.50 |
| 220 | 20-25 | 11.70 |
| 220 | 30 | 12.75 |
| 220 | 40 | 14.25 |
| 220 | 50 | 15.00 |
| 220 | 75 | 16.50 |
| 220 | 100 | 21.00 |
| 440 | 7½-10-15 | 8.25 |
| 440 | 20-25 | 10.50 |
| 440 | 30 | 11.25 |
| 440 | 40-50 | 12.00 |
| 440 | 75 | 14.25 |
| 440 | 100 | 15.75 |

Hours include checking out of barge storage, handling to erection site, mounting compensator, making connections at compensator and motor and testing motor for direction of rotation.

For four-wire, two-phase motors, add 20 percent to above hours.

Hours do not include installation of conduit, pulling of wire or mounting of motors. See respective tables for these time frames.

# DC MOTOR RHEOSTATS AND SWITCHES

HOURS REQUIRED EACH

| STARTING RHEOSTATS AND EXTERNALLY OPERATED SWITCHES | | | | | |
|---|---|---|---|---|---|
| Voltage | Motor Horsepower | Mounting | Voltage | Motor Horsepower | Mounting |
| 115 | ½-¾-1 | 6.90 | 230 | ½-¾-1 | 6.90 |
| 115 | 1½-2-3 | 7.50 | 230 | 1½-2-3-4-5 | 7.50 |
| 115 | 5 | 10.05 | 230 | 7½ | 10.05 |
| 115 | 7½ | 13.05 | 230 | 10 | 11.10 |
| 115 | 10 | 13.80 | 230 | 15 | 13.05 |
| 115 | 15 | 16.05 | 230 | 20 | 14.10 |
| 115 | 20 | 17.05 | 230 | 25 | 16.05 |
| 115 | 25 | 21.45 | 230 | 30-35 | 17.40 |
| — | — | — | 230 | 40 | 18.30 |
| — | — | — | 230 | 50 | 20.70 |

| PUSH-BUTTON CONTROLLED MAGNETIC SWITCHES AND LINE SWITCHES | | | SPEED-REGULATING RHEOSTATS (CONTROLLED BY RESISTANCE IN ARMATURE CIRCUIT) AND EXTERNALLY OPERATED SWITCHES | | |
|---|---|---|---|---|---|
| Voltage | Motor Horsepower | Mounting | Voltage | Motor Horsepower | Mounting |
| 115 | 1-2-3 | 12.00 | 115 | ½-¾-1 | 7.95 |
| 115 | 5 | 14.10 | 115 | 1½-2-3 | 9.00 |
| 115 | 7½-10 | 16.50 | 115 | 5 | 12.00 |
| 230 | 1-2-3-5 | 12.00 | 115 | 7½-10 | 15.00 |
| 230 | 7½-10 | 14.10 | 230 | ½-¾-1 | 7.95 |
| 230 | 15 | 16.05 | 230 | 1½-2-3 | 9.00 |
| — | — | — | 230 | 5 | 9.60 |
| — | — | — | 230 | 7½-10 | 12.00 |

Above hours include:

*Starting Rheostats and Externally Operated Switches*—Mounting and connecting at switches, rheostats and motors. For speed regulator with shunt field weakening instead of plain starting rheostats, add one hour in each case.

*Push-Button Controlled Magnetic Switches and Line Switches*—Mounting and connecting push-button control station and connections at motor.

*Speed-Regulating Rheostats and Externally Operated Switches*—Mounting and making connections at switch, rheostat and motor.

Hours do not include conduit installation, pulling of wire or mounting of motor. See other tables for these time frames.

# MOUNTING MOTORS

**AC, 60-Cycle, 2- and 3-Phase, 220, 440 or 550 Volts**

HOURS REQUIRED EACH

| Horse-power | RPM RATING AND MOUNTING HOURS | | | | | | |
|---|---|---|---|---|---|---|---|
| | 1750 | 1160 | 875 | 700 | 575 | 490 | 420 |
| 1 | 2.91 | 2.91 | 3.64 | 4.37 | 4.37 | 5.09 | 6.55 |
| 1½ | 3.28 | 3.28 | 4.37 | 4.37 | 5.09 | 6.55 | 8.00 |
| 2 | 3.64 | 3.64 | 4.37 | 5.09 | 6.55 | 8.00 | 10.19 |
| 3 | 3.64 | 4.37 | 5.09 | 6.55 | 8.00 | 10.19 | 14.55 |
| 5 | 4.37 | 5.09 | 6.55 | 8.00 | 10.19 | 14.55 | 15.28 |
| 7½ | 5.09 | 6.55 | 8.00 | 10.19 | 14.55 | 15.28 | 18.92 |
| 10 | 6.55 | 8.00 | 10.19 | 14.55 | 15.28 | 18.92 | 20.37 |
| 15 | 8.00 | 10.19 | 14.55 | 15.28 | 18.92 | 20.37 | 21.83 |
| 20 | 10.19 | 11.64 | 15.28 | 18.92 | 20.37 | 21.83 | 22.55 |
| 25 | 11.64 | 15.28 | 18.92 | 20.37 | 21.83 | 22.55 | 37.83 |
| 30 | 14.55 | 18.92 | 20.37 | 21.83 | 22.55 | 37.83 | 37.83 |
| 35 | 15.28 | 20.37 | 21.83 | 22.55 | 33.47 | 37.83 | 43.65 |
| 40 | 18.92 | 20.37 | 21.83 | 22.55 | 37.83 | 37.83 | 43.65 |
| 50 | 21.83 | 24.73 | 24.73 | 33.47 | 37.83 | 43.65 | 43.65 |
| 60 | 24.73 | 29.10 | 33.47 | 37.83 | 43.65 | 43.65 | 55.29 |
| 75 | 29.10 | 34.92 | 37.83 | 43.65 | 55.29 | 55.29 | 73.14 |
| 100 | 34.92 | 37.83 | 43.65 | 55.29 | 73.14 | 73.14 | 80.70 |
| 125 | 34.92 | 42.20 | 55.29 | 73.14 | 73.14 | 80.70 | 89.53 |
| 440- or 550-Volt Motors | | | | | | | |
| 150 | 37.83 | 42.20 | 55.29 | 73.14 | 80.70 | 89.53 | 93.31 |
| 175 | 42.20 | 49.47 | 73.14 | 80.70 | 89.53 | 102.14 | 103.40 |
| 200 | 42.20 | 49.47 | 73.14 | 80.70 | 89.53 | 102.14 | 103.40 |

Above hours include checking out of barge storage, rigging, picking, setting and aligning of motor.

Hours do not include installation of supports, conduit, pulling of wire or connecting. See respective tables for these time frames.

# MOUNTING MOTORS

**Variable Speeds To RPM Rating Listed**
**AC, 60-Cycle, 2- and 3-Phase, 220, 440 or 550 Volts**

HOURS REQUIRED EACH

| Horse-power | RPM RATING AND MOUNTING HOURS | | | | | |
|---|---|---|---|---|---|---|
| | 1750 | 1160 | 875 | 700 | 575 | 490 |
| 1 | 3.64 | 3.64 | 3.64 | 3.64 | 6.55 | 6.55 |
| 1½ | 3.64 | 3.64 | 3.64 | 3.64 | 6.55 | 6.55 |
| 2 | 3.64 | 3.64 | 3.64 | 6.55 | 8.00 | 13.10 |
| 3 | 3.64 | 3.64 | 5.09 | 8.00 | 13.10 | 16.00 |
| 5 | 3.64 | 5.09 | 6.55 | 13.10 | 16.00 | 16.00 |
| 7½ | 5.09 | 6.55 | 8.00 | 16.00 | 16.00 | 20.37 |
| 10 | 6.55 | 8.00 | 13.10 | 16.00 | 20.37 | 21.83 |
| 15 | 8.00 | 13.10 | 16.00 | 20.37 | 21.83 | 23.28 |
| 20 | 13.10 | 16.00 | 16.00 | 21.83 | 23.28 | 26.19 |
| 25 | 13.10 | 16.00 | 20.37 | 23.28 | 24.73 | 34.92 |
| 30 | 16.00 | 20.37 | 21.83 | 24.73 | 34.92 | 39.30 |
| 35 | — | — | 23.28 | 34.92 | 34.92 | 39.30 |
| 40 | 20.37 | 21.83 | 23.28 | 34.92 | 40.74 | 48.02 |
| 50 | 23.28 | 24.73 | 34.92 | 34.92 | 40.74 | 48.02 |
| 60 | 23.28 | 24.73 | 34.92 | 40.74 | 48.02 | 61.11 |
| 75 | 23.28 | 37.83 | 48.02 | 52.96 | 52.96 | 76.92 |
| 100 | 32.01 | 37.83 | 48.02 | 52.96 | 76.92 | 76.92 |
| 440- and 550-Volt Motors | | | | | | |
| 125 | 37.83 | 43.65 | 52.96 | 76.92 | 76.92 | 85.75 |
| 150 | 46.56 | 43.65 | 68.09 | 85.75 | 85.75 | 85.75 |
| 175 | 46.56 | 55.29 | 74.40 | 85.75 | 94.58 | 104.66 |
| 200 | 46.56 | 55.29 | 74.40 | 94.58 | 94.58 | 104.66 |

Above hours include checking out of barge storage, rigging, picking, setting and aligning of motor.

Hours do not include installation of supports, conduit, pulling of wire or connecting. See respective tables for these time frames.

# MOUNTING MOTORS

**Constant and Variable Speeds**
**AC, 25-Cycle, 2- and 3-Phase, 220, 440 or 550 Volts**

HOURS REQUIRED EACH

| Horsepower | RPM RATING AND MOUNTING HOURS | | | | | |
|---|---|---|---|---|---|---|
| | CONSTANT SPEEDS | | | VARIABLE SPEEDS | | |
| | 1440 | 720 | 475 | 1440 | 720 | 475 |
| 1 | 3.28 | 3.64 | 4.37 | 4.37 | 4.37 | 4.37 |
| 1½ | 3.64 | 4.37 | 5.09 | 4.37 | 4.37 | 5.09 |
| 2 | 3.64 | 4.37 | 5.09 | 4.37 | 5.09 | 5.09 |
| 3 | 4.37 | 5.09 | 8.00 | 5.09 | 6.55 | 8.00 |
| 5 | 5.09 | 6.55 | 10.19 | 6.55 | 8.00 | 13.10 |
| 7½ | 8.00 | 8.00 | 14.55 | 8.00 | 13.10 | 16.00 |
| 10 | 10.19 | 10.19 | 15.28 | 13.10 | 16.00 | 21.83 |
| 15 | 10.19 | 15.28 | 20.37 | 16.00 | 21.83 | 23.28 |
| 20 | 14.55 | 18.92 | 21.83 | 21.83 | 24.73 | 24.73 |
| 25 | 14.55 | 20.37 | 22.55 | 21.83 | 24.73 | 37.83 |
| 30 | 20.37 | 21.83 | 34.92 | 23.28 | 24.73 | 42.20 |
| 35 | 21.83 | 22.55 | 37.83 | 24.73 | 37.83 | 42.20 |
| 40 | 21.83 | 22.55 | 39.30 | 24.73 | 37.83 | 42.20 |
| 50 | 29.10 | 34.92 | 48.02 | 37.83 | 42.20 | 52.38 |
| 60 | 29.10 | 39.30 | 48.02 | 39.29 | 52.38 | 66.93 |
| 75 | 29.10 | 39.30 | 48.02 | 42.20 | 52.38 | 66.93 |
| 100 | 37.83 | 46.56 | 61.11 | 52.38 | 58.01 | 87.01 |
| 125 | 46.56 | 78.18 | 87.00 | 52.38 | 81.97 | 99.62 |
| 440- or 550-Volt Motors | | | | | | |
| 150 | 58.20 | 87.00 | 94.58 | 58.01 | 87.01 | 99.62 |
| 175 | 65.96 | 94.58 | 94.58 | 73.14 | 89.53 | 109.71 |
| 200 | 65.96 | 94.58 | 107.19 | 81.97 | 89.53 | 109.71 |

Above hours include checking out of barge storage, rigging, picking, setting and aligning of motor.

Hours do not include installation of supports, conduit, pulling of wire or connecting. See respective tables for these time frames.

# MOUNTING MOTORS

**DC, 115-230 Volts, Constant Speeds, Shunt Series and Compound-Wound Commutating Pole-Type**

HOURS REQUIRED EACH

| Horse-power | RPM RATING AND MOUNTING HOURS | | | | | | |
|---|---|---|---|---|---|---|---|
| | 1750 | 1150 | 850 | 690 | 575 | 500 | 450 |
| 1 | 3.28 | 3.28 | 4.37 | 5.09 | 5.09 | 6.55 | 6.55 |
| 1½ | 4.37 | 4.37 | 5.09 | 5.09 | 5.09 | 6.55 | 6.55 |
| 2 | 4.37 | 4.37 | 5.09 | 6.55 | 6.55 | 10.19 | 10.19 |
| 3 | 4.37 | 5.09 | 6.55 | 10.19 | 10.19 | 10.19 | 11.64 |
| 5 | 5.09 | 6.55 | 10.19 | 10.19 | 11.64 | 14.55 | 14.55 |
| 7½ | 6.55 | 10.19 | 11.64 | 11.64 | 14.55 | 16.00 | 16.00 |
| 10 | 10.19 | 11.64 | 14.55 | 14.55 | 16.00 | 21.83 | 21.83 |
| 15 | 11.64 | 15.28 | 16.00 | 21.83 | 21.83 | 23.28 | 29.10 |
| 20 | 15.28 | 16.00 | 21.83 | 21.83 | 24.73 | 29.10 | 37.83 |
| 25 | 16.00 | 21.83 | 21.83 | 24.73 | 29.10 | 37.83 | 53.84 |
| 30 | 16.00 | 21.83 | 21.83 | 29.10 | 37.83 | 53.84 | 53.84 |
| 40 | 21.83 | 21.83 | 29.10 | 37.83 | 53.84 | 53.84 | 56.75 |
| 50 | 21.83 | 24.73 | 37.83 | 53.84 | 56.75 | 56.75 | 68.09 |
| 60 | 21.83 | 29.10 | 53.84 | 56.75 | 56.75 | 68.09 | 68.09 |
| 75 | 24.73 | 37.83 | 56.75 | 66.93 | 68.09 | 68.09 | 85.75 |
| 100 | 37.83 | 53.84 | 56.75 | 78.57 | 85.75 | 94.58 | 100.88 |
| 125 | 53.84 | 56.75 | 68.09 | 85.75 | 94.58 | 108.45 | 115.11 |
| 150 | 56.75 | 68.09 | 87.00 | 106.80 | 108.45 | 115.11 | 117.28 |
| 175 | 68.09 | 87.00 | 97.10 | 108.45 | 118.53 | 119.80 | 122.32 |
| 200 | 87.00 | 97.10 | 108.45 | 118.53 | 119.80 | 122.32 | 126.10 |

Above hours include checking out of barge storage, rigging, picking, setting and aligning of motor.

Hours do not include installation of supports, conduit, pulling of wire or connecting. See respective tables for these time frames.

# FIRE ALARM SYSTEMS

HOURS REQUIRED EACH

| Item Description | Hours Required |
|---|---|
| Alarm Stations: | |
| Manual | 0.67 |
| Automatic Thermostat | 1.17 |
| Special Outlet Box | 1.04 |
| Alarm Signals: | |
| Bell | 0.91 |
| Flush Horn | 1.17 |
| Megaphone Horn | 1.43 |
| Grille Horn | 1.95 |
| Siren—Small Motor-Driven | 1.43 |
| Chimes | .78 |
| Annunciators—Mounting | 1.63 |
| Annunciators—Per Terminal | .33 |
| Control Panel | 5.20 |
| Auxiliary Station | 2.60 |

Above hours include checking out of barge storage, rigging, picking, setting in position, installing or mounting and make-up or connecting of items as outlined.

If special rough-in can is used rather than a standard outlet box, add 1.5 hours per can.

Hours do not include installation of supports, conduit, or pulling of wire. See respective tables for these time requirements.

# INTERCOMMUNICATION SYSTEMS

HOURS REQUIRED PER ITEM

| Item Description | Hours Required |
|---|---|
| Telephone Systems: | |
| Set in Place and Mount Central Switchboard | 10.50 |
| Central Switchboard Connections—Per Terminal Pair | 0.38 |
| Local Stations: | |
| Flush-Mounted Talk-Back Speaker Stations | 2.25 |
| Surface-Mounted Talk-Back Speaker Stations | 1.50 |
| Suspended Wall or Desk Handsets | 1.50 |
| Cradle Desk Sets | 1.36 |
| Desk-Type Intercom Systems: | |
| Master Station Per Terminal | 0.75 |
| Substations | 1.65 |

# PUBLIC ADDRESS SYSTEMS

HOURS REQUIRED PER ITEM

| Item Description | Hours Required |
|---|---|
| Amplifiers | 5.53 |
| Mounting | 0.33 |
| Connecting—Per Terminal | 0.91 |
| Microphone Outlet | 0.65 |
| Mounting and Connecting Microphones or Speakers: | |
| Flush Wall | 1.95 |
| Flush Ceiling | 2.47 |
| Surface Wall | 1.04 |
| Mounting and Connecting Trumpet-Type Speakers: | |
| Indoor Wall-Mounted | 2.60 |
| Outdoor Wall-Mounted | 3.90 |
| Outdoor Pole-Mounted | 5.20 |

Above hours include checking out of barge storage, rigging, picking, installing or mounting and connecting of items as outlined.

Hours do not include installation of supports, conduit or pulling of wire. See respective tables for these time

# FLUORESCENT AND MERCURY VAPOR LIGHT FIXTURES

### All Metal Open-Type—Pendant-Mounted

HOURS REQUIRED EACH

| Channel | Item Description | Hours Required |
|---|---|---|
| Four Feet | 2-4 Foot Standard Lamps | 2.78 |
| Four Feet | 3-4 Foot Standard Lamps | 3.09 |
| Six Feet | 2-6 Foot Standard Lamps | 3.64 |
| Six Feet | 3-6 Foot Standard Lamps | 3.71 |
| Eight Feet | 2-8 Foot Standard Lamps | 4.25 |
| Eight Feet | 3-8 Foot Standard Lamps | 4.41 |
| Eight Feet | 4-4 Foot Standard Lamps | 4.79 |
| Eight Feet | 6-4 Foot Standard Lamps | 5.80 |

### High-Intensity Mercury Vapor Units

HOURS REQUIRED EACH

| Item Description | Hours Required |
|---|---|
| 400-Watt Fixture | 2.03 |
| Transormer | 2.78 |
| Supporting Brackets | 1.58 |

Above hours include checking out of barge storage, rigging, hoisting and placing of fixture on deck, and fixture installation.

Hours do not include installation of conduit, wire pulling, special fastenings, hanger or suspension means such as long anchors or conduit suspension stems. See other tables for these time requirements.

# INCANDESCENT LIGHTING UNITS

HOURS REQUIRED EACH

| Item Description | Assemble and Hang |
|---|---|
| Ceiling Unit and Glassware: | |
| 100 to 200 Watts | 0.62 |
| 300 to 500 Watts | 0.78 |
| 750 to 1000 Watts | 0.93 |
| Chain or Suspension Unit and Glassware: | |
| 100 to 200 Watts | 0.93 |
| 300 to 500 Watts | 1.08 |
| 750 to 1000 Watts | 1.31 |
| Miscellaneous Ceiling Units: | |
| Flush Units 9″ × 9″ | 2.32 |
| Flush Units 12″ × 12″ | 3.09 |
| Outlet Box Lamp Receptacle | 0.40 |
| Drop Cords | 0.56 |
| Lamp Guards | — |
| Wall Fixtures: | |
| Exit Lights—Surface Type | 0.93 |
| Small Fixture, Indoor or Outdoor | 0.62 |
| Base Light | 1.55 |
| Reel Extension Light | 1.55 |
| Reel Power Extension Fixture | 9.30 |

Above hours include checking out of barge storage, rigging, hoisting and placing of fixture on deck and fixture installation.

Hours do not include installation of conduit, wire pulling, hook-up or lamp installation. See respective tables for these time frames.

# LAMP INSTALLATION
## Incandescent and Fluorescent

HOURS REQUIRED EACH

| Item Description | Hours Required |
|---|---|
| Incandescent Lamps: | |
| 25 Watts | 0.10 |
| 40 Watts | 0.10 |
| 60 Watts | 0.10 |
| 75 Watts | 0.10 |
| 100 Watts | 0.10 |
| 150 Watts | 0.10 |
| 200 Watts | 0.12 |
| 300 Watts | 0.12 |
| 500 Watts | 0.12 |
| 750 Watts | 0.13 |
| 1000 Watts | 0.13 |
| Fluorescent Lamps: | |
| 4-feet | 0.24 |
| 6-feet | 0.40 |
| 8-feet | 0.56 |

Above hours include checking out of barge storage, handling to erection site and installing lamp in fixture.

Hours do not include installation of conduit, fixtures, switches or wire pulling. See other tables for these time frames.

# OUTDOOR FLOOD LIGHTING
## Assemble and Install

HOURS REQUIRED EACH

| Item Description | Hours Required |
|---|---|
| Flood Lighting: | |
| Open Type | |
| 100 Watts | 3.00 |
| 150 Watts | 3.00 |
| 200 Watts | 3.00 |
| 300 Watts | 3.75 |
| 500 Watts | 3.75 |
| 750 Watts | 7.50 |
| 1000 Watts | 7.50 |
| 1500 Watts | 7.50 |
| Louvers or Visors | 1.13 |
| Steel Pole and Single-Flood | — |
| Steel Pole and Double-Flood | — |
| Enclosed Type: | |
| 200/250 Watts | 6.00 |
| 300/500 Watts | 7.50 |
| 750/1000 Watts | 11.25 |
| 1500 Watts | 11.25 |
| Louvers or Visors | 1.13 |
| Navigation Aids—120-Volt, Single-Phase | 6.00 |

Above hours for flood lighting include assembling, setting and connecting of items as outlined.

Above hours for navigation aids includes installation and connecting of pre-assembled unit.

All the above hours include the obtaining of the individual items from barge storage and handling to the erection site.

Hours do not include installation of supports, conduit or wire pulling. See respective tables for these time frames.

# INSTALLATION OF CABLE TRAY AND FITTINGS

HOURS REQUIRED PER UNITS LISTED

| Tray Item Description | Unit | WIDTH OF TRAY IN INCHES | | | | | | |
|---|---|---|---|---|---|---|---|---|
| | | 6 | 9 | 12 | 18 | 24 | 30 | 36 |
| Ladder Type Cable Tray—Straight | lf. | 0.25 | 0.30 | 0.33 | 0.35 | 0.40 | 0.45 | 0.55 |
| 90° Horizontal Elbows—12" Radius | ea. | 1.25 | 1.25 | 1.50 | 1.90 | 2.50 | 3.00 | 3.50 |
| 90" Vertical Elbows—12" Radius | ea. | 2.19 | 2.19 | 2.63 | 3.33 | 4.38 | 4.98 | 5.80 |
| Horizontal Tees—12" Radius | ea. | 2.30 | 2.30 | 2.75 | 3.50 | 4.60 | 5.25 | 6.10 |
| Horizontal Crosses—12" Radius | ea. | 3.00 | 3.00 | 3.60 | 4.55 | 6.00 | 6.85 | 7.95 |
| Reducer | ea. | — | — | 3.00 | 3.50 | 4.00 | 4.50 | 5.00 |
| Expansion Joint | ea. | 2.50 | 3.00 | 4.00 | 4.75 | 5.50 | 6.25 | 7.00 |
| Connector Plates | pr. | 1.00 | 1.00 | 1.00 | 1.00 | 1.00 | 1.00 | 1.00 |
| Dropouts | ea. | 1.25 | 1.25 | 1.50 | 1.75 | 2.00 | 2.50 | 3.00 |
| Blind Ends | ea. | 0.50 | 0.50 | 1.00 | 1.00 | 1.25 | 1.50 | 1.75 |
| Tray Cover Plate | lf. | 0.10 | 0.12 | 0.15 | 0.20 | 0.25 | 0.50 | 0.75 |
| Cable Separators | ea. | 1.00 | 1.00 | 1.00 | 1.00 | 1.00 | 1.00 | 1.00 |

Above hours are for installlation of ladder type cable tray and fittings with 3⅝" siderails and rungs on 6-inch centers all of 16-gauge steel.

Hours include rigging, picking, setting on platform, cutting, assembling and placing.

Hours do not include structural supports on which cable tray is installed. See respective table for this time frame.

# INSTRUMENTS AND CONTROLS
## Large Case Pneumatic Instruments

HOURS REQUIRED EACH

| Instrument Description | Board-Mounted | | Field-Mounted |
|---|---|---|---|
| | Install | Connect | Install |
| Recorders (14⅜" × 17"): | | | |
| Flow | 10.0 | 7.0 | 20.0 |
| Level | 8.0 | 7.0 | — |
| Pressure | 8.0 | 7.0 | 15.0 |
| Temperature | 10.0 | 7.0 | 25.0 |
| For Each Additional Element/Pen, Add | — | 2.5 | 3.0 |
| For Explosion-Proof Construction, Add | — | 3.0 | 3.0 |
| Recorder-Controllers (14⅜" × 17"): | | | |
| Flow | 10.0 | 8.0 | — |
| Level | 10.0 | 8.0 | — |
| Pressure | 10.0 | 8.0 | 20.0 |
| Temperature | 10.0 | 8.0 | 30.0 |
| For Each Additional Element/Pen, Add | — | 2.5 | 3.0 |
| For Explosion-Proof Construction, Add | — | 3.0 | 3.0 |
| Subpanel With Auto./Man. Service Switch | — | 4.0 | — |
| Subpanel With Auto./Man. Test Service Switch | — | 3.0 | 3.0 |
| Indicating Controller (9" × 11¾"): | | | |
| Level | — | — | 20.0 |
| Pressure | — | — | 20.0 |
| Temperature | — | — | 30.0 |
| For Each Element/Pen, Add | — | — | 3.0 |
| For Explosion-Proof Construction, Add | — | — | 3.0 |

Above hours include checking out of storage, handling, installing, calibrating and testing of instruments as listed.

Should instruments come premounted on panel board, disregard the board-mounted installation hours above and apply the connect hours only.

Hours do not include the installation of pipe, valves, fittings, cable tray or supports. See other tables for these time frames.

# INSTRUMENTS AND CONTROLS
## Circular Case Pneumatic Recorders

HOURS REQUIRED EACH

| Instrument Description | Board-Mounted | |
|---|---|---|
| | Install | Connect |
| Recorders-Circular Case: | | |
| Flow | | |
| Circular Case 12″—1 Pen | 20.0 | 5.0 |
| Circular Case 12″—2 Pen | 25.0 | 6.0 |
| Circular Case 12″—3 Pen | 30.0 | 8.0 |
| Add For Explosion-Proof Construction | 3.0 | 3.0 |
| Pressure (with 13 Different Elements, 0-80,000 PSI) | | |
| Circular Case 12″—1 Pen | 10.0 | 5.0 |
| Circular Case 12″—2 Pen | 12.0 | 7.0 |
| Add For Explosion-Proof Construction | 3.0 | 3.0 |
| Temperature | | |
| Circular Case 12″—1 Pen | 10.0 | 5.0 |
| Circular Case 12″—2 Pen | 12.0 | 7.0 |
| Capillary Tubing With Element Add For Each 10′ | 1.5 | — |
| Add For Explosion-Proof Construction | 3.0 | — |

Above hours include checking out of storage, handling, installing, calibrating and testing of instruments as listed.

Should instruments come premounted on panel board, disregard the board-mounted installation hours above and apply the connect hours only.

Hours do not include the installation of pipe, valves, fitting, cable tray or supports. See other tables for these time frames.

# INSTRUMENT AND CONTROLS
## Pneumatic Local Indicators—Transmitters

HOURS REQUIRED EACH

| Instrument Description | Field Installed |
|---|---|
| Flow: | |
| Integral Bellows, 150 PSI Max., 9″ × 11⅜″ | 30.0 |
| Integral Bellows, 2000 PSI Max., 9″ × 11⅜″ | 25.0 |
| Level and Pressure: | |
| 10 Misc. Elements, 0-6000 PSI, 5¾″ × 8″ | 20.0 |
| Misc. Elements Low & High-Pressure 9″ × 11⅜″ | 20.0 |
| Blind 8″ × 5⅜″ | 12.0 |
| Temperature: | |
| Misc. Systems 5¾″ × 8″ | 20.0 |
| Misc. Systems 9″ × 11⅜″ | 20.0 |
| Blind 8″ × 5¾″ | 15.0 |

Above hours include checking out of storage, handling, installing calibrating and testing of instruments as listed.

Hours do not include the installation of pipe, valves, fittings, cable tray or supports. See other tables for these time frames.

# INSTRUMENTS AND CONTROLS
## Pneumatic Transmitters—D/P Type

HOURS REQUIRED EACH

| Instrument Description | Field Installed |
|---|---|
| Flow: | |
| Maximum 500 PSI | 12.0 |
| Maximum 1500 PSI | 12.0 |
| Maximum 6000 PSI | 15.0 |
| Target Type—Flanged Line Size ¾" | 6.0 |
| Target Type—Weld-in Line Size 2" | 10.0 |
| Target Type—Weld-in Line Size 3" | 12.0 |
| Target Type—Weld-in Line Size 4" | 15.0 |
| Level: | |
| Flat Flange Mounting, Maximum 150 lbs. | 7.0 |
| Flat Flange Mounting, Maximum 300 lbs. | 9.0 |
| Flange With Dome, Maximum 150 lbs. | 9.0 |
| Flange With Dome, Maximum 300 lbs. | 11.0 |
| Pressure: | |
| Absolute | 10.0 |
| Gauge Ranges, Maximum 60 #-3000 #-12000 # | 10.0 |
| Differential | 12.0 |
| Temperature: | |
| Misc. System With Capillary Tube | 12.0 |

Above hours include checking out of storage, handling, installing, calibrating and testing of instruments as listed, including weld-in or bolt-up as may be required.

Hours do not include the installation of pipe, valves, fittings, cable trays or supports. See other tables for these time frames.

# INSTRUMENTS AND CONTROLS
## Large Case (14⅜″ × 17″) Electronic Instruments

HOURS REQUIRED EACH

| Instrument Description | Board Mounted | |
|---|---|---|
| | Install | Connect |
| Dynalog Recorder 115-V 50-W Circ. 12″ Chart | 10.0 | 6.0 |
| Dynalog Indicator Concentric | 8.0 | 5.0 |
| Dynalog Indicator Eccentric | 8.0 | 5.0 |
| For Explosion-Proof Construction, Add | 2.0 | — |
| For Explosion-Proof Chart Drive Only, Add | 2.0 | — |
| For Each Additional Pen/Element, Add | 3.0 | — |
| Resistance Temp. Dynalog Recorder 115-V 35-W Circ. 12″ Chart | 10.0 | 6.0 |
| For Explosion-Proof Construction, Add | 3.0 | 1.0 |
| For Pneumatic Chart Drive, Add | — | 2.0 |
| For Explosion-Proof Chart Drive, Add | — | 3.0 |
| For Each Additional Pen/Element, Add | — | 3.0 |
| Dynalog Receiver Recorder (Mag. Flow) | 10.0 | 6.0 |
| For Explosion-Proof Construction, Add | — | 3.0 |
| For Explosion-Proof Chart Drive Only, Add | — | 2.0 |
| Single-Record Strip Chart Recorder—17¾″ × 19¾″ | 12.0 | 6.0 |
| Dual-Record Strip Chart Recorder—17¾″ × 19¾″ | 15.0 | 10.0 |
| Multi-Point Recorder (2 Points Basic)—17¾″ × 19⅝″ | 12.0 | 8.0 |
| For Each (24 Max.) Additional Point, Add | — | 2.0 |
| Circular Chart Recorders—Controllers 12″—17⅛″ × 16⅝″ | 10.0 | 6.0 |
| Multi-point Indicator (Built in Switches) (48 Points) 17⅛″ × 16⅝″ | 45.0 | 30.0 |
| Multi-point Indicator (Built in Switches) (60 Points) 17⅛″ × 16⅝″ | 55.0 | 40.0 |

Above hours include checking out of storage, handling, installing, hooking-up, calibrating and testing of instruments listed.

Hours do not include installation of pipe, valves, fittings, wire, conduit, cable tray or supports. See other tables for these time frames.

# INSTRUMENTS AND CONTROLS
## Electronic Instruments

HOURS REQUIRED EACH

MAGNETIC FLOW TRANSMITTERS
FLANGED ENDS—CAST ALUMINUM HOUSING—EXPLOSION-PROOF

| Line Size Inches | Size in Inches Length × Width × Height | Parallel Coils Watts | Series Coils Watts | Volts | Field Installed |
|---|---|---|---|---|---|
| ½ | 20 × 11 × 15¼ | — | 23 | 115 | 3.0 |
| 1 | 20 × 11 × 15¼ | 54 | 14 | 115 | 3.0 |
| 1½ | 20 × 11 × 15¼ | 60 | 15 | 115 | 3.5 |
| 2 | 20 × 11-⁹/₁₆ × 15¼ | 72 | 18 | 115 | 4.3 |
| 3 | 24 × 14⅞ × 18¼ | 99 | 23 | 115 | 5.7 |
| 4 | 24 × 14⅞ × 18¼ | 122 | 32 | 115 | 8.0 |
| 6 | 36 × 20 × 22¼ | 135 | 32 | 115 | 10.0 |
| 8 | 36 × 20 × 22¼ | 203 | 50 | 115 | 12.0 |
| 10 | 48 × 28 × 28 | 207 | 68 | 230 | 15.5 |
| 12 | 48 × 31 × 31 | 338 | 100 | 230 | 18.7 |
| 14 | 66 × 37 × 46 | 320 | 80 | 230 | 21.0 |
| 16 | 72 × 41 × 51 | 400 | 100 | 230 | 24.5 |
| 18 | 72 × 44 × 54 | 600 | 150 | 230 | 28.5 |
| 20 | 72 × 47 × 57 | 800 | 200 | 230 | 32.0 |
| 24 | 72 × 55 × 65 | 1000 | 250 | 230 | 36.0 |

Above hours include checking out of storage, handling, installing, hooking-up calibrating and testing of instruments as listed.

Hours do not include installation of pipe, valves, fittings, wire, conduit, cable tray or supports. See other tables for these time frames.

# INSTRUMENTS AND CONTROLS
## Electronic Local Transmitters and Indicators

HOURS REQUIRED EACH

| Instrument Description | Field Installed |
|---|---|
| **Flow:** | |
| Transmitter—Explosion-Proof 500 PSI Max., 65-V DC—6 Terminals | 12.0 |
| Transmitter—Explosion-Proof 1500 PSI Max., 65-V DC—6 Terminals, Dual Range | 12.0 |
| Transmitter—Explosion-Proof, 6000 PSI Max., 65-V DC—6 Terminals, Dual Range | 15.0 |
| Telemeter Transmitter—Element-Mounted, Local Indicator, 9⅛" × 11⅜" | 28.0 |
| Targetflow Transmitter—Flow Range 0-11.5; 0-27¾" Flg.—6 Terminals | 6.0 |
| Targetflow Transmitter—Flow Range 0-28; 0-125—2" Weld-In | 10.0 |
| Targetflow Transmitter—Flow Range 0-60; 0-250—3" Weld-In | 12.0 |
| Targetflow Transmitter—Flow Range 0-150; 0-500—4" Weld-In | 18.0 |
| **Level:** | |
| 150# Flat Flange Mounting—4-¾", 8-¾" or 8-⅞" Bolts—6 Terminals—65-V DC Explosion-Proof | 8.0 |
| 300# Flat Flange Mounting—4-¾", 8-¾" or 8-⅞" Bolts—6 Terminals—65-V DC Explosion-Proof | 10.0 |
| 150# Flange, With Extension—4-¾", 8-¾" or 8-⅞" Bolts—6 Terminals—65-V DC Explosion-Proof | 10.0 |
| 300# Flange With Extension—4-¾", 8-¾" or 8-⅞" Bolts—6 Terminals—65-V DC Explosion-Proof | 12.0 |
| **Pressure:** | |
| Absolute Pressure—2 Ranges, 0-1500 MM Hg, 6 Terminals—65-V DC, Explosion-Proof | 10.0 |
| Differential Pressure—4000 PSI, 6 Terminals—65-V DC, Explosion-Proof | 12.0 |
| Telemeter Transmitter—Local Indicator—115-V, 11 Misc. Elements, 0-6000 PSI | 18.0 |
| Low-Pressure Local Indicator—115-V, Element-Mounting (Vacuum) | 14.0 |
| Differential Transmitter—Local Indicator—115-V, Element-Mounting (Vacuum) | 22.0 |
| Gauge Pressure Transmitter—Max. 350-3000 PSI, 6 Terminals, Explosion-Proof | 12.0 |
| **Temperature:** | |
| Transmitter—Local Indicator, Miscellaneous Bulbs | 12.0 |

Above hours include checking out of storage, handling, installing, hooking-up, calibrating and testing of instruments as listed.

Hours do not include installation of pipe, valves, fittings, wire conduit, cable tray or supports. See other tables for these time frames.

# INSTRUMENTS AND CONTROLS
## Miscellaneous Level and Flow Instruments

HOURS REQUIRED EACH

| Instrument Description | Hours Required |
|---|---|
| Miscellaneous Level—Board Connect: | |
| Level Indicator—6″ Round Case D/P Cell-Mounted—Max. 6000 PSI | 5.0 |
| Level Indicator—6″ Round Case D/P Cell-Mounted—Max. 10000 PSI | 5.0 |
| Level Indicator—3″ Round Case D/P Cell-Mounted—Max. 500 PSI | 5.0 |
| Miscellaneous Level—Field Installed: | |
| Level Indicator—6″ Round Case D/P Cell-Mounted—Max. 6000 PSI | 10.0 |
| Level Indicator—6″ Round Case D/P Cell-Mounted—Max. 10000 PSI | 10.0 |
| Level Gage Glass—12″ Long | 4.0 |
| Level Gage Glass with Cock Valves—12″ Long | 6.0 |
| Level Gage Glass—For Each Additional 12″, Add | 1.0 |
| Level Gage Glass—Transparent with Illuminator—12″ Long | 8.0 |
| Level Control Switch | 3.0 |
| Level Controller Float Operated External Steel Cage | 12.0 |
| Miscellaneous Flow—Field Installed: | |
| Flow Meter, Purge-Rator Type | 3.0 |
| Sight Flow Indicators, Screwed ½″-¾″-1″ | 2.0 |
| Sight Flow Indicators, Screwed 1¼″-1½″ | 2.5 |
| Sight Flow Indicators, Screwed 2″ | 3.2 |
| Sight Flow Indicators, Flanged ½″-¾″-1″ | 3.0 |
| Sight Flow Indicators, Flanged 1¼″-1½″ | 3.5 |
| Sight Flow Indicators, Flanged 2″ | 4.8 |

Above hours include checking out of storage, handling, installing, calibrating and testing of instruments as listed.

Hours do not include the installation of pipe, valves, fittings, conduit, wire cable tray or supports. See respective tables for these time frames.

# INSTRUMENTS AND CONTROLS
## Miscellaneous In-Line Instruments

HOURS REQUIRED EACH

| Line Size | PRESSURE RATING | | | | | |
|---|---|---|---|---|---|---|
| | 150 lb. | 300 lb. | 400 lb. | 600 lb. | 900 lb. | 1500 lb. |
| 1 | 1.7 | 1.9 | — | 2.4 | — | 3.0 |
| 1½ | 1.8 | 2.0 | — | 2.8 | — | 3.6 |
| 2 | 1.9 | 2.4 | — | 3.1 | — | 3.9 |
| 3 | 2.8 | 3.3 | — | 4.0 | 4.0 | 5.4 |
| 4 | 4.1 | 4.8 | 4.8 | 5.6 | 5.6 | 7.3 |
| 6 | 5.2 | 6.1 | 6.1 | 7.8 | 7.8 | 9.3 |
| 8 | 7.0 | 8.2 | 8.2 | 9.4 | 9.4 | 12.7 |
| 10 | 9.0 | 10.2 | 10.2 | 11.5 | 11.5 | — |
| 12 | 11.1 | 12.7 | 12.7 | 14.5 | 14.5 | — |
| 14 | 12.7 | 14.6 | 14.6 | — | — | — |
| 16 | 14.7 | 16.9 | 16.9 | — | — | — |
| 18 | 16.3 | 18.9 | 18.9 | — | — | — |
| 20 | 18.7 | 21.6 | 21.6 | — | — | — |
| 24 | 21.7 | 25.1 | — | — | — | — |
| 30 | 27.2 | — | — | — | — | — |
| 36 | 32.6 | — | — | — | — | — |

There are a variety of instruments which, when installed, actually become a part of the pipeline in which they are installed. The above hours take into consideration the installation of the instrument, including two flange-ups, and checking out of storage, handling, calibrating when necessary, and testing.

Hours do not include the installation of pipe, valves, fittings, conduit, wire, cable tray or supports. See other tables for these time frames.

# INSTRUMENTS AND CONTROLS
## Miscellaneous Temperature, Pressure and Other Instruments

HOURS REQUIRED EACH

| Instrument Description | Hours Required |
|---|---|
| Miscellaneous Temperature: | |
| Bi-Metal Thermometer | 2.0 |
| Thermowell | 2.0 |
| Thermowell With Chain and Cap | 2.0 |
| Thermocouple Assembly—Screwed | 3.0 |
| Thermocouple Assembly—Flanged | 4.0 |
| Thermocouple Assembly—Flanged R.J. | 4.5 |
| Temperature Switch With Capillary Tube—Explosion-Proof Mercoid | 3.5 |
| Miscellaneous Pressure: | |
| Pressure Gage—4½" Dial | 3.0 |
| Draft Gage | 4.0 |
| Pressure Switch—Explosion-Proof | 3.0 |
| Miscellaneous—Miscellaneous: | |
| Siphon | 0.5 |
| Valve Positioners | 2.5 |
| Integrally Mounted Valve Positioner | 4.0 |
| Valve Operator | 5.0 |
| Air Filters | 1.5 |
| Air Pressure Regulator | 1.0 |
| Air Pressure-Regulator (Combination) | 1.5 |
| Adjustable Restrictor (Damper) | 2.0 |
| Alarms (Panel-Mounted, Single) | 1.5 |
| Alarms (Panel-Mounted, Dual) | 2.0 |
| Vibration Switches—Electric | 4.0 |

Above hours include checking out of storage, handling, installing, and calibrating and testing of instrument if necessary.

Hours do not include the installation of pipe, valves, fittings, conduit, wire, cable tray or supports. See other tables for these time frames.

# INSTRUMENTS AND CONTROLS
## Installation of Multi-Tube Bundles

HOURS REQUIRED PER LINEAR FOOT

| Tube Type | Size Inches | NUMBER OF TUBES IN BUNDLE | | | | | | | | | | |
|---|---|---|---|---|---|---|---|---|---|---|---|---|
| | | 2 | 3 | 4 | 5 | 7 | 8 | 10 | 12 | 14 | 19 | 37 |
| Copper With | ¼ | .04 | .05 | .06 | .07 | .09 | .10 | .12 | .14 | .16 | .22 | .32 |
| Plastic Sheath | ⅜ | .05 | .06 | .07 | .08 | .11 | .14 | .18 | .22 | — | — | — |
| only | ½ | .07 | .09 | .13 | — | — | — | — | — | — | — | — |
| Copper With | ¼ | .05 | .06 | .07 | .08 | .10 | .12 | .14 | .16 | .18 | .25 | .37 |
| Armour & Plastic | ⅜ | .06 | .07 | .08 | .09 | .12 | .16 | .20 | .25 | — | — | — |
| Sheath | ½ | .08 | .10 | .14 | — | — | — | — | — | — | — | — |
| Aluminum With | — | — | — | — | — | — | — | — | — | — | — | — |
| Plastic Sheath | ¼ | — | — | .05 | .06 | .08 | .08 | .09 | .11 | .12 | .18 | .25 |
| only | — | — | — | — | — | — | — | — | — | — | — | — |
| Aluminum With | — | — | — | — | — | — | — | — | — | — | — | — |
| Armour & Plastic | ¼ | — | — | .06 | .07 | .09 | .09 | .10 | .13 | .14 | .20 | .28 |
| Sheath | — | — | — | — | — | — | — | — | — | — | — | — |
| Plastic Mylar | — | — | — | — | — | — | — | — | — | — | — | — |
| Envel. & Vinyl | ¼ | .03 | .04 | .05 | .06 | .07 | .08 | .09 | .10 | .12 | .14 | .20 |
| Jacket | ⅜ | .04 | .04 | .06 | .07 | .09 | .09 | .11 | .12 | — | — | — |

| Fitting Type | Hours Each |
|---|---|
| Standard Indoor Junction Box 12″ × 18″ × 6″ | 2.5 |
| Weather Tight Junction Box 12″ × 18″ × 6″ | 4.0 |
| Union Box 10″ × 5″ × 5″ | 2.0 |
| Neoprene Grommets | 0.3 |
| Cast Aluminum Connector | 0.5 |
| Galvanized Connector and Neoprene Weather Proof Bushing | 0.9 |

Above hours include checking out of storage, handling to erection location, and complete installation with average time allowed for make-up.

Hours do not include installation of instruments, cable tray or supports. See respective tables for these time frames.

# INSTRUMENTS AND CONTROLS
## Installation of Single Tubing, Fittings and Valves

HOURS REQUIRED PER UNITS LISTED

| Material Item | SIZE IN INCHES | | | | | |
|---|---|---|---|---|---|---|
| | Unit | ¼ | ⅜ | ½ | ¾ | 1 |
| Copper Tubing | lf. | .30 | .35 | .40 | .46 | .52 |
| Aluminum Tubing | lf. | .30 | .35 | .40 | .46 | .52 |
| Stainless Steel Tubing | lf. | .60 | .65 | .75 | .82 | .90 |
| Plastic Tubing | lf. | .30 | .35 | .40 | .46 | .52 |
| Steel Tubing | lf. | .40 | .48 | .53 | .59 | .65 |
| Fittings Up To 600 lb. | ea. | .40 | .40 | .40 | .50 | .50 |
| Fittings Over 600 lb. | ea. | .40 | .40 | .40 | .50 | .50 |
| Valves—Screwed | ea. | 1.00 | 1.00 | 1.00 | 1.20 | 1.30 |
| Valves—Flanged | ea. | 1.60 | 1.60 | 1.60 | 1.60 | 1.70 |

Above hours include checking out of storage, handling to erection location and complete installation with average time allowed for make-up.

Hours do not include installation of instruments, cable tray or supports. See respective tables for these time frames.

# EQUIPMENT AND APPURTENANCES INSTALLATION

## Derrick Barge Labor Crew

| Personnel Description | Number of Men |
|---|---|
| Superintendent or Captain | 1 |
| Assistant Superintendent | 1 |
| Barge Foreman | 2 |
| Anchor Foreman | 2 |
| Pipefitter Foreman | 2 |
| Welder Foreman | 2 |
| Rigger Foreman | 2 |
| Leaderman | 4 |
| Crane Operator—600 Ton | 2 |
| Crane Operator—100 Ton | 2 |
| Anchor Hoist Operator | 4 |
| Hoist Operator | 4 |
| Oiler | 4 |
| Pipefitters | 4 |
| Pipe And Structural Welders | 12 |
| Instrument Fitters | 4 |
| Electricians | 4 |
| Riggers/Helpers | 20 |
| Iron Workers | 4 |
| Welder Helpers | 12 |
| Millwrights | 4 |
| X-Ray Technician | 2 |
| X-Ray Helper | 2 |
| Project Engineer | 1 |
| Radio Operator/Technician | 1 |
| Clerk | 1 |
| Assistant Clerk/Radio Operator | 1 |
| Total Crew | 104 |

The above labor crew is for the supervision and installation of equipment, prefabricated pipe spools, and electrical and instrument items on a marine structure. This crew can be enlarged or diminished depending on the scope of work to be performed.

In addition to the above personnel the derrick barge maintenance, quartering and catering, tug, crew boat and sand blast and paint crews must be added to obtain a complete working spread. See section six, "Offshore Structures" for these crews.

# EQUIPMENT AND APPURTENANCES INSTALLATION

## Derrick Barge Equipment Spread

| Equipment Description | Number of Units |
|---|---|
| Barge With Revolving—600-Ton Crane | 1 |
| Crawler Crane—600-Ton | 1 |
| Anchor Handling Winches | 8 |
| Anchors—12,000 Pounds Each ± | 8 |
| 1½" Anchor Wire Rope—3000 lin. ft. ea. | 8 |
| Generators, 250-KW | 4 |
| Air Compressor—600 CFM | 2 |
| Welding Machines—400-AMP | 20 |
| X-Ray Equipment (Complete Set-Up) | 1 |
| Water Pumps 4"—6" | 4 |
| Short-Wave Radio System | 1 |

The above equipment spread is for use in the installation of equipment, prefabricated spools, and electrical and instrument items on marine structures. This spread can be added to or diminished depending on the scope of work to be performed.

In addition to the above spread the tug and crew boat spreads, a quartering barge if required, and sandblast and paint equipment must be added to obtain a complete working spread. See section six, "Offshore Structures" for these spreads.

# Section Eight
# MISCELLANEOUS CONSTRUCTION ITEMS

In preparing an estimate for pipelines and offshore structures, a variety of direct cost items must be considered and added to the estimate if necessary.

This section presents these items as a descriptive checklist. Any or all of the items may be needed, depending on the requirements of the project being considered.

## Mobilization and Demobilization of Personnel

This involves getting all the personnel required to construct the project to and from the site, should they not be available at the site. Consideration should be given the following items if they are required:

1) *Processing*—The cost of hiring people through the personnel office. If the project is in a country foreign to the personnel being hired the process cost of obtaining passports, visas, and necessary medical inoculations should be charged as a part of this operation. If contract personnel are to be employed, include the cost of preparing contracts. Any cost of foreign clearance should also be included.
2) *Travel Cost*—The round-trip cost of all airplane, train, or bus fares and any other travel costs should be charged here.
3) *Travel Pay*—The straight time wage rate for an eight-hour day should be paid each individual for each day spent traveling and charged here.
4) *Per diem*—A set daily amount to be paid the individual for meals, lodging en route if necessary, and miscellaneous items while traveling.

## Mobilization and Demobilization of Equipment

Like labor, equipment must be moved to and from the job site. The following should be considered and charged under this item:

1) *Rig-Up And Rig-Down of Construction Equipment*—The rigging up of construction equipment with all required components and items as may be required to perform the work for which it is intended and the disassembling of same after construction completion should be charged under this heading. This work can be accomplished either before or after the equipment is shipped to the construction site, depending on the type of equipment involved.
2) *Equipment Load-Out*—All cost incurred for loading out and preparing the equipment for shipment to the job site.
3) *Transport Construction Equipment*—The time required to move the construction equipment from a set location to the project location round-trip. This equipment should be charged to the job at the non-working day rate for the total transit time.
4) *Labor*—Labor required to move the equipment from one location to another at the established wage rates. In the case of mobilizing marine equipment, fully manned tugs will be required and should be charged at the agreed daily rate. If a lay/derrick barge is to be mobilized, a riding or maintenance crew will be required and should be included at the agreed daily rate. See page 303 for suggested riding crew. Processing of all personnel will be required as listed on page 300.
5) *Permits And Fees*—The cost of all road or travel permits, wharfage fees, pilot fees, duties, etc., that may be required to move the equipment from one location to another.

## Load-Out, Tie-Down and Material Handling

Items that should be considered under this heading are packaging or crating if required, loading out all required materials on carriers and securing for shipment.

1) *Packaging and Crating*—The cost of labor and materials required to package or crate items prior to shipment. If this work is to be performed as a subcontract, the cost of the subcontract should be charged under this heading.
2) *Load-Out*—The cost of all labor and equipment required to load the materials and processing equipment or structures onto carriers for shipment to the construction site.
3) *Tie-Down*—The cost of all labor and miscellaneous supplies required for tying down or securing the permanent materials, equipment or structures on carriers for shipment to the construction site. See table on page 303 for suggested labor crew for this operation.
4) *Carrier Equipment*—The daily cost of tractor-trailers, rail cars, ships or barges during the load-out period. If the carrier is to remain at the construction site as a part of the construction equipment or for storage, its time and charge should be assessed under mobilization and demobilization while traveling and directly against the project upon its arrival at the work site. If its use is soley for transporting permanent materials to the job site, unloading, and returning to its point of origin, its cost should be assessed as a direct job cost.
5) *Permits, Liscenses, Fees and Duties*—Any charges assessed against the materials to be transported, such as load-out yard fees, wharfage fees, pilot fees, export-import duties, etc. These charges are not to be confused with those listed under "Mobilization and Demobilization of Equipment."

## Transportation and Freight

The delivery cost of items added to the project after work has begun, late delivery items, shortage replacements or items that were missed when compiling the estimate should be included under this heading. Usually an allowance is included in the estimate, based on past experience, for this item. In any event, this is in addition to and not to be confused with Mobilization and Demobilization, or Load-Out, Tie-Down and Material Handling. The following items should be considered:

1) *Freight Delivery*—The cost of transporting the materials by truck, rail, ship or air.
2) *Packaging*—The total cost of packaging, crating or making the materials ready for shipment.
3) *Load-Out*—The cost of loading and securing the materials on the type of transportation selected.
4) *Taxes, Fees, Duties*—The cost of any taxes, export-import duties, wharfage and pilot fees or custom duties.

## Camp, Family Quarters, Field Office, and Storage Facilities

Bachelor camp facilities, family living quarters, field office, warehousing, and storage yards should be considered under this heading. For cross-country pipelines, more than one complete set-up may be required depending on length of line and location. For near-shore marine pipelines or structures, it may be desirable to quarter and cater personnel and control the project from facilities located on shore. If the project location is at sea, a considerable distance from shore, quartering and catering facilities aboard the lay/derrick barge or a special barge rigged for this purpose should be utilized and charged to the project at the agreed daily rate. The estimator should give consideration to the following if any or part of these facilities are required:

1) *Camp Facilities*—Complete living quarters, including ample facilities for personal hygene, recreation, and eating for the number of desired personnel. All necessary kitchen facilities, cooking utensils, dishes, silverware, bed linens and towels and necessary personnel to maintain, operate and manage the camp should be included. If desired, a total installation, maintenance and operating cost for camp facilities can usually be obtained from an organization engaged in this type of business.
2) *Family Status*—Rental houses, contractor constructed houses or mobile homes may be used if providing family living facilities is a requirement.
3) *Field Office, Warehousing and Storage*—Contractor constructed or rented structures or trailers and fenced yards in the size and quantity as may be required to support the project.

## Rest and Recreation

When a project is confined to an isolated area and long working hours are a requisite, it is common to work a two-week shift and allow one week off for rest and recreation. Should this be the case, all cost incurred should be charged under this item. Things to consider are:

1) *Transportation Cost*—The round-trip cost of air, bus, rail or ship travel from the construction site to the rest and recreation area.
2) *Per diem Pay*—The cost which may be incurred for lodging, food and miscellaneous incidentals while traveling.
3) *Quartering and Catering*—The cost of lodging and meals during the rest and recreation period. Depending on the agreement this item may or may not be a part of the conditions.
4) *Family Status*—If the project includes facilities and allows the employee to be accompanied to the project with his family, the above may apply to the family as well as the employee, depending on the conditions of the contract.

# LAY OR DERRICK BARGE RIDING CREW

| Personnel Description | Number of Men |
|---|---|
| Barge Superintendent or Captain | 1 |
| Chief Engineer | 1 |
| Shift Engineer | 2 |
| Electrician | 1 |
| Operator | 1 |
| Crewman | 2 |
| Cook | 1 |
| Mess Boy | 2 |
| Janitor | 1 |
| Total Crew | 12 |

The above crew is sufficient as a riding crew to maintain and operate the lay or derrick barge when moving from one location to another.

# LOAD-OUT AND TIE-DOWN CREW

| Personnel Description | Number of Men |
|---|---|
| Superintendent | 1 |
| Barge Foreman | 1 |
| Welding Foreman | 1 |
| Welders | 12 |
| Welder Helpers | 12 |
| Crane Operator | 4 |
| Hoist Operator | 2 |
| Riggers/Laborers | 20 |
| Clerk | 2 |
| Total Crew | 55 |

The above crew is ample for loading-out and tying-down of pipe, jackets, platforms and equipment of considerable tonnage for offshore installation.

The project at hand should be evaluated as to the amount of materials to be shipped and time allowed to load-out and tie-down same. The above crew can be adjusted upward or downward as necessary.

# Section Nine
# MISCELLANEOUS ESTIMATING INFORMATION

As stated in the Preface, this manual is solely for the estimation of time frames, labor crews, and equipment spreads required for the installation of land and sea pipelines, marine facilities, and offshore structures and is not intended for their design. Therefore, this section has been held to a minimum and includes only information that we feel will be beneficial to the estimator in preparation of his estimate.

Included are tables showing, pipe and tubular sizes, wall thicknesses, and weights, welding rod requirements for pipe and tubulars, surface area and circumferences of pipe for computing painting and welding time, pipe coating and wrapping material requirements, useful conversion factors, mensuration units, and a conversion table converting minutes to decimal hours.

# PIPE WALL THICKNESS AND WEIGHT

POUNDS PER LINEAR FOOT

| Schedule Number | Nominal Size | 2″ | 2½″ | 3″ | 3½″ | 4″ | 5″ | 6″ | 8″ | 10″ | 12″ |
|---|---|---|---|---|---|---|---|---|---|---|---|
| | O. D. | 2.375 | 2.875 | 3.5 | 4.0 | 4.5 | 5.563 | 6.625 | 8.625 | 10.75 | 12.75 |
| | Wall Thickness | — | — | — | — | — | — | — | — | — | — |
| 10 | Weight/Foot | — | — | — | — | — | — | — | — | — | — |
| | Wall Thickness | — | — | — | — | — | — | — | .250 | .250 | .250 |
| 20 | Weight/Foot | — | — | — | — | — | — | — | 22.36 | 28.04 | 33.38 |
| | Wall Thickness | — | — | — | — | — | — | — | .277 | .307 | .330 |
| 30 | Weight/Foot | — | — | — | — | — | — | — | 24.70 | 34.24 | 43.77 |
| | Wall Thickness | .154 | .203 | .216 | .226 | .237 | .258 | .280 | .322 | .365 | .406 |
| 40 | Weight/Foot | 3.65 | 5.79 | 7.59 | 9.11 | 10.79 | 14.62 | 18.97 | 28.55 | 40.48 | 53.52 |
| | Wall Thickness | — | — | — | — | — | — | — | .406 | .500 | .562 |
| 60 | Weight/Foot | — | — | — | — | — | — | — | 35.64 | 54.75 | 73.15 |
| | Wall Thickness | .218 | .276 | .300 | .318 | .337 | .375 | 4.32 | .500 | .594 | .688 |
| 80 | Weight/Foot | 5.02 | 7.66 | 10.25 | 12.51 | 14.98 | 20.78 | 28.57 | 43.39 | 64.43 | 88.63 |
| | Wall Thickness | — | — | — | — | — | — | — | .594 | .719 | .844 |
| 100 | Weight/Foot | — | — | — | — | — | — | — | 50.95 | 77.03 | 107.32 |
| | Wall Thickness | — | — | — | — | .438 | .500 | .562 | .719 | .844 | 1.000 |
| 120 | Weight/Foot | — | — | — | — | 19.00 | 27.04 | 36.39 | 60.71 | 89.29 | 125.49 |
| | Wall Thickness | — | — | — | — | — | — | — | .812 | 1.000 | 1.125 |
| 140 | Weight/Foot | — | — | — | — | — | — | — | 67.76 | 104.13 | 139.67 |
| | Wall Thickness | .344 | .375 | .438 | — | .531 | .625 | .719 | .906 | 1.125 | 1.312 |
| 160 | Weight/Foot | 7.46 | 10.01 | 14.32 | — | 22.51 | 32.96 | 45.35 | 74.69 | 115.64 | 160.27 |
| Double | Wall Thickness | .436 | .552 | .600 | .636 | .674 | .750 | .864 | .875 | — | — |
| Extra Heavy | Weight/Foot | 9.03 | 13.70 | 18.58 | 22.85 | 27.54 | 38.55 | 53.16 | 72.42 | — | — |
| Non- | Wall Thickness | — | — | .156 | — | .156 | .188 | .188 | .188 | .188 | .188 |
| Scheduled | Weight/Foot | — | — | 5.57 | — | 7.24 | 10.79 | 12.92 | 16.94 | 21.21 | 25.22 |
| Non- | Wall Thickness | — | — | .188 | — | .188 | — | .250 | .219 | .203 | .375 |
| Scheduled | Weight/Foot | — | — | 6.65 | — | 8.66 | — | 17.02 | 19.66 | 22.87 | 49.56 |
| Non- | Wall Thickness | — | — | — | — | — | — | .375 | .375 | .279 | .500 |
| Scheduled | Weight/Foot | — | — | — | — | — | — | 25.03 | 33.04 | 31.20 | 65.42 |

# PIPE WALL THICKNESS AND WEIGHT

POUNDS PER LINEAR FOOT

| Schedule Number | Nominal Size | 14″ | 16″ | 18″ | 20″ | 22″ | 24″ | 26″ | 30″ | 36″ |
|---|---|---|---|---|---|---|---|---|---|---|
| | O.D. | 14.0 | 16.0 | 18.0 | 20.0 | 22.0 | 24.0 | 26.0 | 30.0 | 36.0 |
| 10 | Wall Thickness | .250 | .250 | .250 | .250 | .250 | .250 | .312 | .312 | .312 |
| | Weight/Foot | 36.71 | 42.05 | 47.39 | 52.73 | 58.07 | 63.41 | 85.60 | 98.93 | 118.92 |
| 20 | Wall Thickness | .312 | .312 | .312 | 3.75 | .375 | .375 | .500 | .500 | .500 |
| | Weight/Foot | 45.61 | 52.27 | 58.94 | 78.60 | 86.61 | 94.62 | 136.17 | 157.53 | 189.57 |
| 30 | Wall Thickness | .375 | .375 | .438 | .500 | .500 | .562 | — | .625 | .625 |
| | Weight/Foot | 54.57 | 62.58 | 82.15 | 104.13 | 114.81 | 140.68 | — | 196.08 | 236.13 |
| 40 | Wall Thickness | .438 | .500 | .562 | .594 | — | .688 | — | — | .750 |
| | Weight/Foot | 63.44 | 82.77 | 104.67 | 123.11 | — | 171.29 | — | — | 282.35 |
| 60 | Wall Thickness | .594 | .656 | .750 | .812 | .875 | .969 | — | — | — |
| | Weight/Foot | 85.05 | 170.50 | 138.17 | 166.40 | 197.41 | 238.35 | — | — | — |
| 80 | Wall Thickness | .750 | .844 | .938 | 1.031 | 1.125 | 1.219 | — | — | — |
| | Weight/Foot | 106.13 | 136.61 | 170.92 | 208.87 | 250.81 | 296.58 | — | — | — |
| 100 | Wall Thickness | .938 | 1.031 | 1.156 | 1.281 | 1.375 | 1.531 | — | — | — |
| | Weight/Foot | 130.85 | 164.82 | 207.96 | 256.10 | 302.88 | 367.45 | — | — | — |
| 120 | Wall Thickness | 1.094 | 1.219 | 1.375 | 1.500 | 1.625 | 1.812 | — | — | — |
| | Weight/Foot | 150.79 | 192.43 | 244.14 | 296.37 | 353.61 | 429.50 | — | — | — |
| 140 | Wall Thickness | 1.250 | 1.438 | 1.562 | 1.750 | 1.875 | 2.062 | — | — | — |
| | Weight/Foot | 170.21 | 223.64 | 274.22 | 341.10 | 403.01 | 483.24 | — | — | — |
| 160 | Wall Thickness | 1.406 | 1.594 | 1.781 | 1.969 | 2.125 | 2.344 | — | — | — |
| | Weight/Foot | 189.11 | 245.25 | 308.55 | 379.14 | 451.07 | 542.09 | — | — | — |
| Non- | Wall Thickness | .281 | .281 | .281 | .281 | .281 | .281 | .250 | .250 | .250 |
| Scheduled | Weight/Foot | 41.17 | 47.17 | 53.18 | 59.18 | 65.18 | 71.18 | 68.75 | 79.43 | 95.45 |
| Non- | Wall Thickness | .500 | .438 | .375 | .625 | .625 | .500 | .375 | .375 | .375 |
| Scheduled | Weight/Foot | 72.09 | 72.80 | 70.59 | 129.33 | 142.68 | 125.49 | 102.63 | 118.635 | 142.68 |
| Non- | Wall Thickness | — | .750 | .500 | .750 | — | .750 | .625 | .750 | — |
| Scheduled | Weight/Foot | — | 122.15 | 93.45 | 154.19 | — | 186.23 | 169.38 | 234.29 | |

# TUBULAR SIZE, WALL THICKNESS AND WEIGHT

POUNDS PER LINEAR FOOT

| Nominal Size Inches | WALL THICKNESS IN INCHES | | | | | | | |
|---|---|---|---|---|---|---|---|---|
| | .250 | .375 | .500 | .625 | .750 | .875 | 1.000 | 1.125 |
| 24 | 63.41 | 94.62 | 125.49 | 156.05 | 186.23 | 216.13 | 245.67 | 274.88 |
| 25 | 66.09 | 98.64 | 130.85 | 162.73 | 194.27 | 225.48 | 256.35 | 286.90 |
| 26 | 68.75 | 102.63 | 136.17 | 169.38 | 202.28 | 234.82 | 267.04 | 298.91 |
| 27 | 71.43 | 106.65 | 141.53 | 176.08 | 210.29 | 244.17 | 277.72 | 310.93 |
| 28 | 74.10 | 110.65 | 146.87 | 182.75 | 218.30 | 253.52 | 288.40 | 322.95 |
| 29 | 76.77 | 114.66 | 152.21 | 189.43 | 226.31 | 262.86 | 299.08 | 334.96 |
| 30 | 79.43 | 118.65 | 157.53 | 196.08 | 234.29 | 272.21 | 309.76 | 346.98 |
| 31 | 82.11 | 122.67 | 162.89 | 202.78 | 242.34 | 281.56 | 320.44 | 359.00 |
| 32 | 84.78 | 126.68 | 168.23 | 209.46 | 250.35 | 290.90 | 331.12 | 371.01 |
| 33 | 87.45 | 130.68 | 173.57 | 216.13 | 258.36 | 300.25 | 341.81 | 383.03 |
| 34 | 90.12 | 134.69 | 178.91 | 222.81 | 266.37 | 309.59 | 352.49 | 395.05 |
| 35 | 92.80 | 138.69 | 184.25 | 229.48 | 274.38 | 318.94 | 303.17 | 407.06 |
| 36 | 95.45 | 142.68 | 189.57 | 236.13 | 282.37 | 328.29 | 373.85 | 419.08 |
| 37 | 98.14 | 146.70 | 194.94 | 242.84 | 290.40 | 337.63 | 384.53 | 431.10 |
| 38 | 100.81 | 150.71 | 200.28 | 249.51 | 298.41 | 346.98 | 395.21 | 443.11 |
| 39 | 103.48 | 154.71 | 205.62 | 256.19 | 306.42 | 356.33 | 405.89 | 455.13 |
| 40 | 106.15 | 158.72 | 210.96 | 262.86 | 314.43 | 365.67 | 416.58 | 467.15 |
| 41 | 108.82 | 162.73 | 216.30 | 269.54 | 322.45 | 375.02 | 427.26 | 479.16 |
| 42 | 111.49 | 166.73 | 221.64 | 276.22 | 330.46 | 384.36 | 437.94 | 491.18 |
| 43 | 114.16 | 170.74 | 226.98 | 282.89 | 338.47 | 393.71 | 448.62 | 503.20 |
| 44 | 116.83 | 174.74 | 232.32 | 289.57 | 346.48 | 403.06 | 459.30 | 515.21 |
| 45 | 119.50 | 178.75 | 237.66 | 296.24 | 354.49 | 412.40 | 469.98 | 527.23 |
| 46 | 122.17 | 182.75 | 243.00 | 302.92 | 362.50 | 421.75 | 480.66 | 539.25 |
| 47 | 124.84 | 186.76 | 248.34 | 309.59 | 370.51 | 431.10 | 491.35 | 551.26 |
| 48 | 127.51 | 190.76 | 253.68 | 316.27 | 378.52 | 440.44 | 502.03 | 563.28 |
| 49 | 130.18 | 194.77 | 259.02 | 322.95 | 386.53 | 449.79 | 512.71 | 575.30 |
| 50 | 132.85 | 198.77 | 264.37 | 329.62 | 394.55 | 459.14 | 523.39 | 587.31 |
| 51 | 135.52 | 202.78 | 269.71 | 336.30 | 402.56 | 468.48 | 534.07 | 599.33 |
| 52 | 138.19 | 206.79 | 275.05 | 342.97 | 410.57 | 477.83 | 544.75 | 611.35 |
| 53 | 140.86 | 210.79 | 280.39 | 349.65 | 418.58 | 487.17 | 555.43 | 623.36 |
| 54 | 143.53 | 214.80 | 285.73 | 356.33 | 426.59 | 496.52 | 566.12 | 635.38 |
| 55 | 146.20 | 218.80 | 291.07 | 363.00 | 434.60 | 505.87 | 576.80 | 647.40 |
| 56 | 148.87 | 222.81 | 296.41 | 369.68 | 442.61 | 515.21 | 587.48 | 659.41 |
| 57 | 151.54 | 226.81 | 301.75 | 376.35 | 450.62 | 524.56 | 598.16 | 671.43 |
| 58 | 154.21 | 230.82 | 307.09 | 383.03 | 458.63 | 533.91 | 608.84 | 683.45 |
| 59 | 156.88 | 234.82 | 312.43 | 389.71 | 466.65 | 543.25 | 619.52 | 695.46 |
| 60 | 159.55 | 238.83 | 317.77 | 396.38 | 474.66 | 552.60 | 630.20 | 707.48 |
| 61 | 162.22 | 242.84 | 323.11 | 403.06 | 482.67 | 561.94 | 640.89 | 719.50 |
| 62 | 164.89 | 246.84 | 328.45 | 409.73 | 490.68 | 571.29 | 651.57 | 731.51 |

# TUBULAR SIZE, WALL THICKNESS AND WEIGHT

POUNDS PER LINEAR FOOT

| Nominal Size Inches | WALL THICKNESS IN INCHES | | | | | | | |
|---|---|---|---|---|---|---|---|---|
| | 1.250 | 1.375 | 1.500 | 1.625 | 1.750 | 1.875 | 2.000 | 2.125 |
| 24 | 303.75 | 332.29 | 360.50 | 388.37 | 415.91 | 443.11 | 469.98 | 496.52 |
| 25 | 317.11 | 346.98 | 376.52 | 405.73 | 434.60 | 463.14 | 491.35 | 519.22 |
| 26 | 330.46 | 361.67 | 392.54 | 423.09 | 453.09 | 483.17 | 512.71 | 541.92 |
| 27 | 343.81 | 376.35 | 408.56 | 440.44 | 471.99 | 503.20 | 534.07 | 564.61 |
| 28 | 357.16 | 391.04 | 424.59 | 457.80 | 490.68 | 523.22 | 555.43 | 587.31 |
| 29 | 370.51 | 405.73 | 440.61 | 475.16 | 509.37 | 543.25 | 576.30 | 610.01 |
| 30 | 383.86 | 420.41 | 456.63 | 492.51 | 528.06 | 563.28 | 598.16 | 632.71 |
| 31 | 397.22 | 435.10 | 472.65 | 509.87 | 546.76 | 583.31 | 619.52 | 655.41 |
| 32 | 410.57 | 449.79 | 488.67 | 527.23 | 565.45 | 603.33 | 640.89 | 678.10 |
| 33 | 423.92 | 464.48 | 504.70 | 544.59 | 584.14 | 623.36 | 662.25 | 700.80 |
| 34 | 437.27 | 479.16 | 520.72 | 561.94 | 602.83 | 643.39 | 683.61 | 723.50 |
| 35 | 450.62 | 493.85 | 536.75 | 579.30 | 621.53 | 663.42 | 704.98 | 746.20 |
| 36 | 463.98 | 508.54 | 552.76 | 596.66 | 640.22 | 683.45 | 726.34 | 768.90 |
| 37 | 477.33 | 523.22 | 568.78 | 614.02 | 658.91 | 703.47 | 747.70 | 791.59 |
| 38 | 490.68 | 537.91 | 584.81 | 631.37 | 677.60 | 725.50 | 769.06 | 814.29 |
| 39 | 504.03 | 552.60 | 600.83 | 648.73 | 696.30 | 743.53 | 790.43 | 838.99 |
| 40 | 517.38 | 567.28 | 616.85 | 666.09 | 714.99 | 763.56 | 811.79 | 859.16 |
| 41 | 530.73 | 581.97 | 632.88 | 683.45 | 733.68 | 783.58 | 833.15 | 882.39 |
| 42 | 544.09 | 596.66 | 648.90 | 700.80 | 752.37 | 803.61 | 854.52 | 905.09 |
| 43 | 557.44 | 611.35 | 664.92 | 718.16 | 771.07 | 823.64 | 875.88 | 927.78 |
| 44 | 570.79 | 626.03 | 680.94 | 735.52 | 789.76 | 843.67 | 897.24 | 950.48 |
| 45 | 584.14 | 640.72 | 696.96 | 752.87 | 808.95 | 863.69 | 918.60 | 973.18 |
| 46 | 597.49 | 655.41 | 712.88 | 770.23 | 827.14 | 883.72 | 939.97 | 995.88 |
| 47 | 610.84 | 670.09 | 729.01 | 787.59 | 845.84 | 903.75 | 961.33 | 1018.58 |
| 48 | 624.20 | 684.78 | 745.03 | 804.95 | 864.53 | 923.78 | 982.69 | 1041.27 |
| 49 | 637.55 | 699.47 | 761.05 | 822.30 | 883.22 | 948.81 | 1004.06 | 1063.97 |
| 50 | 650.90 | 714.15 | 777.07 | 839.66 | 901.91 | 963.83 | 1025.42 | 1086.67 |
| 51 | 664.25 | 728.84 | 793.10 | 857.02 | 920.61 | 983.86 | 1046.78 | 1109.37 |
| 52 | 677.60 | 743.53 | 809.12 | 874.38 | 939.30 | 1003.89 | 1068.14 | 1132.07 |
| 53 | 690.96 | 758.22 | 825.14 | 891.73 | 957.99 | 1023.92 | 1089.51 | 1154.76 |
| 54 | 704.31 | 772.90 | 841.16 | 909.09 | 976.68 | 1043.94 | 1110.87 | 1177.46 |
| 55 | 717.66 | 787.59 | 857.19 | 926.45 | 995.38 | 1063.97 | 1132.23 | 1200.16 |
| 56 | 731.01 | 802.28 | 873.21 | 943.81 | 1014.07 | 1084.00 | 1153.60 | 1222.86 |
| 57 | 744.36 | 816.96 | 889.23 | 961.16 | 1032.76 | 1104.03 | 1174.96 | 1245.56 |
| 58 | 757.71 | 831.65 | 905.25 | 978.52 | 1051.45 | 1124.05 | 1196.32 | 1268.25 |
| 59 | 771.07 | 846.34 | 921.27 | 995.88 | 1070.15 | 1144.08 | 1217.68 | 1290.95 |
| 60 | 784,42 | 861.02 | 937.30 | 1013.23 | 1088.84 | 1164.11 | 1239.05 | 1313.65 |
| 61 | 797.77 | 875.71 | 953.32 | 1030.59 | 1107.53 | 1184.14 | 1260.41 | 1336.35 |
| 62 | 811.12 | 890.40 | 963.34 | 1047.95 | 1126.22 | 1204.17 | 1281.77 | 1359.05 |

# WELDING ROD REQUIREMENTS

WALL THICKNESS AND POUNDS OF WELD ROD PER JOINT

| Nominal Pipe Size Inches | PIPE SCHEDULE NUMBERS 10 | | 20 | | 30 | | STD. | | 40 | | 60 | |
|---|---|---|---|---|---|---|---|---|---|---|---|---|
| | A | B | A | B | A | B | A | B | A | B | A | B |
| 2 | .109 | .019 | — | — | — | — | .154 | .058 | .154 | .058 | — | — |
| 3 | .120 | .025 | — | — | — | — | .216 | .126 | .216 | .126 | — | — |
| 4 | .120 | .037 | — | — | — | — | .237 | .190 | .237 | .190 | — | — |
| 6 | .134 | .168 | — | — | — | — | .280 | .669 | .280 | .669 | — | — |
| 8 | .148 | .216 | .250 | .356 | .227 | .860 | .322 | .860 | .322 | .860 | .406 | 1.580 |
| 10 | .165 | .250 | .250 | .411 | .307 | .613 | .365 | .993 | .365 | .993 | .500 | 1.830 |
| 12 | .180 | .320 | .250 | .526 | .330 | 1.270 | .375 | 1.270 | .406 | 2.340 | .562 | 3.440 |
| 14 | .250 | .580 | .312 | .140 | .375 | 1.400 | .375 | 1.400 | .438 | 2.580 | .593 | 3.790 |
| 16 | .250 | .662 | .312 | .160 | .375 | 1.600 | .375 | 1.600 | .500 | 2.950 | .653 | 6.470 |
| 18 | .250 | .744 | .312 | .178 | .438 | 3.310 | .375 | 1.780 | .562 | 3.310 | .750 | 7.230 |
| 20 | .250 | .828 | .375 | 2.000 | .500 | 3.680 | .375 | 2.000 | .593 | 6.290 | .812 | 8.910 |
| 22 | .250 | .910 | .375 | 2.200 | .500 | 4.050 | .375 | 2.200 | — | — | .875 | 11.460 |
| 24 | .250 | .992 | .375 | 2.400 | — | — | .375 | 2.400 | .687 | 7.540 | .968 | 15.860 |
| 26 | .312 | 1.610 | .500 | 4.790 | — | — | .375 | 2.600 | — | — | .625 | 7.030 |
| 28 | .312 | 2.800 | .500 | 5.150 | .625 | 7.560 | .375 | 2.800 | — | — | 1.000 | 18.520 |
| 30 | .312 | 3.000 | .500 | 5.520 | .625 | 8.100 | .375 | 3.000 | — | — | 1.000 | 19.830 |
| 32 | .312 | 3.200 | .500 | 5.890 | .625 | 8.650 | .375 | 3.200 | .688 | 10.060 | 1.000 | 21.170 |
| 34 | .312 | 3.400 | .500 | 6.260 | .625 | 9.180 | .375 | 3.400 | .688 | 10.680 | 1.000 | 22.480 |
| 36 | .312 | 3.600 | .500 | 6.620 | .625 | 9.720 | .375 | 3.600 | .750 | 14.450 | 1.000 | 23.790 |
| 38 | .250 | 1.570 | .375 | 3.800 | .500 | 6.990 | .625 | 10.270 | .750 | 15.260 | 1.000 | 25.130 |
| 40 | .250 | 1.650 | .375 | 4.000 | .500 | 7.360 | .625 | 10.810 | .750 | 16.060 | 1.000 | 26.450 |
| 42 | .250 | 1.750 | .625 | 11.340 | .750 | 16.860 | .375 | 4.200 | 1.000 | 27.760 | 1.250 | 41.500 |

*Code:*
A Wall thickness of pipe.
B Pounds of weld metal required per joint.

Above pounds of weld metal required per joint is based on 60° total bevel.

With the exception of standard weight and extra heavy wall 42-inch pipe, the wall thicknesses shown below the ruled line are not regularly scheduled pipe.

# WELDING ROD REQUIREMENTS

WALL THICKNESS AND POUNDS OF WELD ROD PER JOINT

| Nominal Pipe Size Inches | PIPE SCHEDULE NUMBERS | | | | | | | | | |
|---|---|---|---|---|---|---|---|---|---|---|
| | XH | | 80 | | 120 | | 160 | | XXH | |
| | A | B | A | B | A | B | A | B | A | B |
| 2 | .218 | .095 | .218 | .095 | — | — | .343 | .229 | .436 | .470 |
| 3 | .300 | .189 | .300 | .189 | — | — | .438 | .562 | .600 | .826 |
| 4 | .337 | .458 | .337 | .458 | .438 | .844 | .531 | 1.240 | .674 | 1.840 |
| 6 | .432 | 1.230 | .432 | 1.230 | .562 | 1.810 | .718 | 2.680 | .864 | 3.480 |
| 8 | .500 | 1.580 | .500 | 1.580 | .718 | 3.450 | .906 | 5.680 | .875 | 4.480 |
| 10 | .500 | 1.830 | .593 | 2.680 | .843 | 5.170 | 1.125 | 8.100 | — | — |
| 12 | .500 | 2.340 | .843 | 6.630 | 1.000 | 8.410 | 1.312 | 14.940 | — | — |
| 14 | .500 | 2.580 | .750 | 5.630 | 1.093 | 11.440 | 1.406 | 19.340 | — | — |
| 16 | .500 | 2.950 | .843 | 8.340 | 1.218 | 13.060 | 1.593 | 25.560 | — | — |
| 18 | .500 | 3.310 | .937 | 11.900 | 1.375 | 21.130 | 1.781 | 37.490 | — | — |
| 20 | .500 | 3.680 | 1.031 | 21.480 | 1.500 | 27.610 | 1.968 | 43.230 | — | — |
| 22 | .500 | 4.050 | 1.125 | 23.640 | 1.625 | 35.140 | 2.125 | 54.140 | — | — |
| 24 | .500 | 4.410 | 1.218 | 23.710 | 1.872 | 47.230 | 2.343 | 69.080 | — | — |
| 26 | .500 | 4.790 | .750 | 10.450 | 1.875 | 54.210 | 2.000 | 61.150 | — | — |
| 28 | .500 | 5.150 | .750 | 11.240 | 1.875 | 58.350 | 2.000 | 65.820 | — | — |
| 30 | .500 | 5.520 | .750 | 12.040 | 1.875 | 62.490 | 2.000 | 70.490 | — | — |
| 32 | .500 | 5.890 | .750 | 12.850 | 1.875 | 66.700 | 2.000 | 75.250 | — | — |
| 34 | .500 | 6.260 | .750 | 13.650 | 1.875 | 70.840 | 2.000 | 79.920 | — | — |
| 36 | .500 | 6.620 | .750 | 14.454 | 1.875 | 74.980 | 2.000 | 84.590 | — | — |
| 38 | 1.250 | 37.070 | 1.500 | 52.430 | 1.875 | 79.200 | 2.000 | 89.350 | — | — |
| 40 | 1.250 | 39.530 | 1.500 | 55.450 | 1.875 | 83.340 | 2.000 | 94.020 | — | — |
| 42 | .500 | 7.730 | 1.500 | 57.910 | 1.875 | 87.480 | 2.000 | 98.690 | — | — |

*Code:*

A Wall thickness of pipe.

B Pounds of weld metal required per joint.

Above pounds of weld metal required per joint is based on 60° total bevel.

With the exception of standard weight and extra heavy wall 42 inch pipe, the wall thicknesses shown below the ruled line are not regularly scheduled pipe.

# SURFACE AREA AND CIRCUMFERENCES OF PIPE FOR COMPUTING PAINTING AND WELDING

| Nominal Size Inches | Surface Area of Pipe Per Lin. Ft. Sq. Ft./Lin. Ft. (Painting) | Circumferences of Pipe In Inches (Welding) |
|---|---|---|
| 1 | 0.344 | 4.24 |
| 1½ | 0.497 | 6.28 |
| 2 | 0.622 | 8.0 |
| 2½ | 0.753 | 9.0 |
| 3 | 0.916 | 11.0 |
| 3½ | 1.047 | 12.6 |
| 4 | 1.178 | 14.2 |
| 5 | 1.456 | 17.48 |
| 6 | 1.734 | 21.0 |
| 8 | 2.258 | 28.0 |
| 10 | 2.81 | 33.77 |
| 12 | 3.34 | 37.7 |
| 14 | 3.67 | 44.0 |
| 16 | 4.19 | 50.27 |
| 18 | 4.71 | 56.55 |
| 20 | 5.24 | 62.83 |
| 24 | 6.28 | 75.40 |
| 30 | 7.85 | 94.25 |
| 36 | 9.43 | 113.10 |
| 42 | 11.00 | 131.95 |
| 48 | 12.57 | 150.80 |
| 60 | 15.71 | 188.50 |

# COATING AND WRAPPING MATERIALS

AMOUNT OF MATERIALS REQUIRED PER MILE OF PIPE

| Nominal Pipe Size | Primer Gallons | 3/32" Coal-tar Enamel Tons | Glass Wrap Squares | Kraft Paper Squares | 15-Pound Felt Squares |
|---|---|---|---|---|---|
| 4 | 11 | 3 | 75 | 75 | 75 |
| 6 | 17 | 4 | 105 | 105 | 105 |
| 8 | 22 | 6 | 135 | 135 | 135 |
| 10 | 28 | 7 | 165 | 165 | 165 |
| 12 | 33 | 8 | 195 | 195 | 195 |
| 14 | 39 | 10 | 210 | 210 | 210 |
| 16 | 44 | 11 | 240 | 240 | 240 |
| 18 | 50 | 13 | 270 | 270 | 270 |
| 20 | 55 | 14 | 300 | 300 | 300 |
| 24 | 66 | 17 | 360 | 360 | 360 |
| 30 | 83 | 21 | 450 | 450 | 450 |
| 36 | 99 | 25 | 540 | 540 | 540 |
| 42 | 116 | 29 | 630 | 630 | 630 |

*Primer and Coal-Tar Enamel:*

The above gallons of primer and tons of coal-tar enamel are based on covering new pipe in good condition. For rough pipe, increase the above primer quantities by 25 percent and the coal-tar quantities by 20 percent. No allowance is included in the above quantities for spillage or waste.

*Wrappings:*

The above squares are based on machine wrapping, allowing ¾-inch laps on pipe sizes through 10-inch, and 1-inch laps on pipe sizes greater than 10 inches. No allowance is included for wrapping drips, patching holidays or short ends which are usually discarded.

# USEFUL CONVERSION FACTORS

## Linear Measure

| | |
|---|---|
| Centimeter | = 0.3937 inch |
| Chain | = 66.00 feet |
| Chain | = 4.00 rods |
| Foot | = 0.3048 meter |
| Inch | = 2.54 centimeters |
| Meter | = 3.281 feet |
| Meter | = 39.37 inches |
| Mile | = 5280 feet |
| Mile | = 1.609 kilometers |
| Rod | = 16.5 feet |
| Yard | = 0.9144 meter |

## Area Units

| | |
|---|---|
| Acre | = 43,560 square feet |
| Acre | = 4,047 square meters |
| Acre | = 160 square rods |
| Square centimeter | = 0.1550 square inch |
| Square foot | = 0.0929 square meter |
| Square inch | = 6.452 square centimeters |
| Square kilometer | = 0.3861 square mile |
| Square meter | = 10.76 square feet |
| Square mile | = 2.590 square kilometers |

## Cubic measure (Volume)

| | |
|---|---|
| Barrel | = 5.6146 cubic feet |
| Barrel | = 42.0 gallons |
| Cubic centimeter | = 0.06102 cubic inch |
| Cubic foot | = 0.1781 barrel |
| Cubic foot | = 7.4805 gallons (U.S.) |
| Cubic foot | = 0.02832 cubic meter |
| Cubic inch | = 16.387 cubic centimeters |
| Cubic meter | = 6.2897 barrels |
| Cubic meter | = 35.314 cubic feet |
| Cubic meter | = 1.308 cubic yards |
| Cubic yard | = 4.8089 barrels |
| Cubic yard | = 46,656 cubic inches |
| Cubic yard | = 0.7646 cubic meter |
| Gallon (U.S.) | = 0.02381 barrel |
| Gallon (U.S.) | = 0.1337 cubic feet |
| Gallon (U.S.) | = 231.0 cubic inches |
| Gallon (U.S.) | = 3.785 liters |
| Gallon (U.S.) | = 0.8327 gallon (Imperial) |
| Liter | = 0.2642 gallon |
| Liter | = 1.0567 quarts |
| Quart (liquid) | = 0.946 liter |

## Weight

| | |
|---|---|
| Grain (Avoirdupois) | = 0.06480 gram |
| Gram | = 15.432 grains |
| Gram | = 0.03527 ounce |
| Kilogram | = 2.2046 pounds |
| Ounce (Avoirdupois) | = 437.5 grains |
| Ounce (Avoirdupois) | = 28.3495 grams |
| Pound | = 0.4536 kilogram |
| Ton (long) | = 2,240 pounds |
| Ton (metric) | = 2,205 pounds |
| Ton (short) | = 2,000 pounds |

## Pressure

| | |
|---|---|
| Atmosphere | = 33.94 feet of water |
| Atmosphere | = 29.92 inches of mercury |
| Atmosphere | = 760.0 millimeters of mercury |
| Atmosphere | = 14.7 pounds per square inch |
| Foot of water at 60°F. | = 0.4331 pound per square inch |
| Inch of mercury | = 1.134 feet of water |
| Inch of mercury | = 0.4912 pound per square inch |
| Inch of water at 60°F. | = 0.0361 pound per square inch |
| Pound per square inch | = 2.309 feet of water at 60°F. |
| Pound per square inch | = 2.0353 inches of mercury |
| Pound per square inch | = 51.697 millimeters of mercury |

## Flow Rates

| | |
|---|---|
| Barrel per hour | = 0.0936 cubic feet per minute |
| Barrel per hour | = 0.700 gallon per minute |
| Barrel per hour | = 2.695 cubic inches per second |
| Barrel per day | = 0.02917 gallon per minute |
| Cubic foot per minute | = 10.686 barrels per hour |
| Cubic foot per minute | = 28.800 cubic inches per second |
| Cubic foot per minute | = 7.481 gallons per minute |
| Gallon per minute | = 1.429 barrels per hour |
| Gallon per minute | = 0.1337 cubic feet per minute |
| Gallon per minute | = 34.286 barrels per day |

## Velocity

| | |
|---|---|
| Foot per second | = 0.68182 mile per hour |
| Mile per hour | = 1.4667 feet per second |

## Power, Work, Energy and Temperature

| | |
|---|---|
| British Thermal Unit | = 0.2520 kilogram calorie |
| British Thermal Unit | = 0.2928 watt hour |
| B.T.U. per minute | = 0.02356 horsepower |
| Foot pound | = 0.001286 British Thermal Unit |
| Foot pound per second | = 0.001818 horsepower |
| Horsepower | = 42.44 B.T.U.'s per minute |
| Horsepower | = 33,000 foot pounds per minute |
| Horsepower | = 550 foot pounds per second |
| Horsepower | = 1.014 horsepower (metric) |
| Horsepower | = 0.7457 kilowatt |
| Horsepower-hour | = 2,547 British Thermal Units |
| Kilogram Calorie | = 3.968 British Thermal Units |
| Kilowatt | = 1.341 horsepower |
| Temp. Centigrade | = 5/9 (Temp °F. − 32) Subtract 32 from the temperature in °F. and multiply the result by 5/9 |
| Temp. Fahrenheit | = 9/5 (Temp. °C.) + 32 Multiply the temperature in C. by 9/5 and then add 32 to the result |
| Watt-hour | = 3.415 British Thermal Units |

# MENSURATION UNITS

Diameter of a circle × 3.1416 = Circumference
Radius of a circle × 6.283185 = Circumference

Square of the radius of a circle × 3.1416 = Area
Square of the diameter of a circle × 0.7854 = Area
Square of the circumference of a circle × 0.07985 = Area
Half the circumference of a circle × half its diameter = Area

Circumference of a circle × 0.159155 = Radius
Square root of the area of a circle × 0.56419 = Radius

Circumference of a circle × 0.31831 = Diameter
Square root of the area of a circle × 1.12838 = Diameter

Diameter of a circle × 0.86 = Side of an inscribed equilateral triangle
Diameter of a circle × 0.7071 = Side of an inscribed square
Circumference of a circle × 0.225 = Side of an inscribed square
Circumference of a circle × 0.282 = Side of an equal square
Diameter of a circle × 0.8862 = Side of an equal square

Base of a triangle × ½ the altitude = Area
Multiplying both diameters and .7854 together = Area of an Elipse

Surface of sphere × 1/6 of its diameter = Solidity

Circumference of a sphere × its diameter = Surface
Square of the diameter of a sphere × 3.1416 = Surface
Square of the circumference of a sphere × 0.3183 = Surface

Cube of the diameter of a sphere × 0.5236 = Solidity
Cube of the radius of a sphere × 4.1888 = Solidity
Cube of the circumference of a sphere × 0.016887 = Solidity

Square root of the surface of a sphere × 0.56419 = Diameter
Square root of the surface of a sphere × 1.772454 = Circumference

Cube root of the solidity of a sphere × 1.2407 = Diameter
Cube root of the solidity of a sphere × 3.8978 = Circumference

Radius of a sphere × 1.1547 = Side of inscribed cube
Square root of (⅓ of the square of) the diameter of a sphere = Side of inscribed cube

Area of its base × ⅓ of its altitude = Solidity of a cone or pyramid, whether round, square or triangular
Area of one of its side × 6 = Surface of cube
Altitude of trapezoid × ½ the sum of its parallel sides = Area

# MINUTES TO DECIMAL HOURS CONVERSION TABLE

| Minutes | Hours | Minutes | Hours |
|---|---|---|---|
| 1 | .017 | 31 | .517 |
| 2 | .034 | 32 | .534 |
| 3 | .050 | 33 | .550 |
| 4 | .067 | 34 | .567 |
| 5 | .084 | 35 | .584 |
| 6 | .100 | 36 | .600 |
| 7 | .117 | 37 | .617 |
| 8 | .135 | 38 | .634 |
| 9 | .150 | 39 | .650 |
| 10 | .167 | 40 | .667 |
| 11 | .184 | 41 | .684 |
| 12 | .200 | 42 | .700 |
| 13 | .217 | 43 | .717 |
| 14 | .232 | 44 | .734 |
| 15 | .250 | 45 | .750 |
| 16 | .267 | 46 | .767 |
| 17 | .284 | 47 | .784 |
| 18 | .300 | 48 | .800 |
| 19 | .317 | 49 | .817 |
| 20 | .334 | 50 | .834 |
| 21 | .350 | 51 | .850 |
| 22 | .368 | 52 | .867 |
| 23 | .384 | 53 | .884 |
| 24 | .400 | 54 | .900 |
| 25 | .417 | 55 | .917 |
| 26 | .434 | 56 | .934 |
| 27 | .450 | 57 | .950 |
| 28 | .467 | 58 | .967 |
| 29 | .484 | 59 | .984 |
| 30 | .500 | 60 | 1.000 |

Lightning Source UK Ltd.
Milton Keynes UK
UKOW030422150513

210695UK00003B/82/P